UNITED NATIONS ECONOMIC COMMISSION FOR EUROPE

TOWARDS A KNOWLEDGE-BASED ECONOMY

REGIONAL ASSESSMENT REPORT

UNITED NATIONS

New York and Geneva, 2002

ECE/TRADE/311/1

UN2
E/ECE/TRADE/311/1

UNITED NATIONS PUBLICATIONS
Sales No. E.03.II.E.9
ISBN 92-1-116823-6

FOREWORD

The last decades of the 20th century have represented a turning point in the global development process. It is knowledge that has become the engine of the social, economic and cultural development in the today's world. Knowledge-intensive economic activities are now a factor of production of strategic importance in the leading countries. They have also become the main indicator of the level of development and the readiness of every country for a further economic and cultural growth in the 21st century. Taking into consideration all these factors, the United Nations Economic Commission for Europe has launched an initiative of monitoring and analyzing the development of the knowledge-based economy in all the European countries in transition and emerging market economies.

The major goal of this initiative is to stimulate the exchange of national experiences, to identify best practices and to promote region-wide and global-wide cooperation of the UNECE member States, which would accelerate the development of a knowledge-based economy in the countries in transition and emerging market economies. It envisages the preparation of country assessment reports on the biennium basis by national experts, nominated by the Governments, the creation of a High-Level Task Force on the Knowledge-Based Economy, which will consider the reports and provide policy advice and recommendations to the participating countries, and the development of progress measurements and indicators, policy guidelines and tools to assist countries in overcoming obstacles to the development of a knowledge-based economy.

We hope that the country assessment reports, showing a detailed level of the countries' potential and providing information on various approaches and solutions, will help policy-makers to take strategic decisions with regards to the challenges facing them in the development of institutions, information and innovation systems, human resources development and other areas crucial for the development of a knowledge-based economy.

Brigita Schmőgnerová
Executive Secretary
United Nations Economic Commission for Europe

PREFACE

The industrial revolution of the 19th century and the scientific revolution of the 20th century have prepared the conditions for the rise of the knowledge-based economy. Economic activities associated with the production and utilization of information and knowledge have become an engine of economic growth in the developed market economies, increasingly transforming all the other dimensions of development and the entire societal *modus vivendi* and *modus operanti* of the humanity.

What do we mean by "the knowledge-based economy"?

It is not just the digital economy, which incorporates the production and use of computers and telecommunication equipment. It is not quite the networked economy, which incorporates the telecommunication and networking growth during the last decades and its impact on human progress.

The knowledge-based economy is a much complex and broader phenomenon. There are different dimensions and aspects of the knowledge-based economy:

1. The knowledge-based economy has a very powerful technological driving force – a rapid growth of information and telecommunication technologies (ICT). Every three – four year there appears a new generation of ICT. Today, the ICT companies are among the largest corporations. The ICT sector is among the fastest growing economic sectors.

2. Telecommunication and networking, stimulated by a rapid growth of ICTs, have penetrated all the spheres of human activity, forcing them to work into an absolutely new mode and creating new spheres. The information society has become a reality.

3. Knowledge, based on information and supported by cultural and spiritual values, has become an independent force and the most decisive factor of social, economic, technological and cultural transformation.

4. The knowledge-based economy has allowed a quick integration of the enormous intellectual resources of economies in transition into the European intellectual pool, stimulating the development of the former countries. Every country can benefit from developing a knowledge-based economy to become a more equal participant in the global development process.

5. The emerging knowledge-based economy has been affecting other areas of societal activity in every country, including institutional and innovation system, human resources development and etc. and visa versa. The knowledge-based economy has become an engine of progress in every country. If a country is developed, it has a developed knowledge-based economy, if a country is lagging behind, a knowledge-based economy constitutes just a small fraction of its economy.

The report below was prepared by a national expert, nominated by the Government, and represents an overview of the present situation and an assessment of the emerging trends in all the major areas, constituting the foundation of the knowledge-based economy, such as policy and policy instruments, institutional regime, ICT infrastructure, information system, national innovation capacities and capabilities.

Introduction

The second half of the 20th century has seen the emergence of a knowledge-based economy that has grown rapidly and has dramatically altered all areas of human activity. By historic standards, there has been a relatively rapid proliferation and diffusion of ICTs. The uptake of ICTs has led to enhanced networking capabilities that, in turn, have given rise to new socio-economic phenomena, including: Networked Access; Networked Learning; Networked Society; Networked Economy; and Networked Policy. This report has analysed these five independent and mutually reinforcing blocks of "Networked" activities. The five areas were assessed, in part, utilising the 19 different categories of knowledge-based economy indicators proposed by the Centre for International Development (CID) at Harvard University. These indicators were presented in the guide "Capture the Benefits of the Networked World" (see: www.readinessguide.org).

The major findings of the report confirm the existence of digital gaps and divides throughout the UNECE region and within member countries.

Since the beginning of the 1990s, digital gaps have been growing within and between the developed market economies and countries in transition. Digital gaps have also appeared within the member-countries of the former USSR, and between the CIS and the Eastern European countries.

Several countries, notably Hungary, Slovenia, and Estonia, are moving rapidly towards a knowledge-based economy and have utilised well the advantages stemming from globalisation and cross-border cooperation.

Certain other countries, such as the Russian Federation, Ukraine and the Republic of Belarus, have been slow to capitalise on their competitive advantages such as human resources, research and development potential, and knowledge stock accumulated over several decades. A lack of investment coupled with monopoly control of the ICT infrastructure has restricted the development of a knowledge-based economy in these countries.

In a number of transition countries, the framework conditions for a knowledge-based economy are either missing or are at an early stage of development. Poverty, for example, was found to be one of the most important factors undermining the development of a knowledge-based economy in many countries in transition.

At the same time, the findings of the report suggest that digital gaps and divides are not irreversible. The Baltic States demonstrate that, under appropriate supporting conditions, these divides can be easily overcome.

The report found that all the transition and emerging market economies of the UNECE region can be placed into three relatively defined subgroups. However, even within these subgroups, variances do exist such that some countries in a less advanced sub-group may have better indicators in specific aspects of the knowledge-based economy than countries in a more advanced subgroup. The report also found that, in some countries, there is such strong potential for dynamic change that the report's findings could rapidly become obsolete.

Structurally, the report follows the outline put forward by the UNECE for country assessments, suggesting that for a knowledge-based economy to emerge the following components must be in place:

- availability and accessibility of information and communication technologies, and an information infrastructure and ICT service-provision;
- a regulatory framework, which is conducive to acquisition, generation and effective utilisation of knowledge and information;
- an active role by government in promoting a community-wide use of knowledge and information and ensuring their effective utilisation and universal access by the population.;
- the availability and efficient utilisation of scientific capacities and capabilities
- an educated and entrepreneurial population.

This summary report utilises the information and data from the nine country assessment reports (including: Armenia; Bulgaria; Georgia; Kyrgyz Republic; Latvia; Republic of Belarus; Russian Federation; Slovakia; and Yugoslavia,) as well as those from other respected sources. The summary report seeks to makes a country comparison utilising the following 19 basic indicators proposed by Harvard University:

Networked Access
- Information infrastructure
- Internet availability
- Internet affordability
- Network speed and quality
- Hardware and software
- Service and support

Networked Learning
- Schools' access to ICTs.
- Enhancing education with ICTs
- Developing the ICT workforce

Networked Society
- People and organisation online
- Locally relevant content
- ICTs in everyday life
- ICTs in the workplace

Networked Economy
- ICT employment opportunities
- B2C electronic commerce
- B2B electronic commerce
- E-Government

Networked Policy
- Telecommunications regulation
- ICT trade policy

1. Information and Telecommunication Technologies: overview and status

1.1. The definition of ICTs

Even among analysts and experts, the term 'Information and Communications Technologies' (ICTs) is viewed and interpreted in different ways. As the name suggests, ICTs encompass all technologies that facilitate the processing and transfer of information and communication services. Some experts also include in the definition all ICT services which are made available with the support of these technologies, and even those human activities which are in any way affected by ICTs. Indeed, rapid advances in information technologies have changed traditional ways of processing information, of conducting communications and of making services available. For example, advances in open systems technology have led to the convergence of communication services, business services, and information services onto one platform, while through unified communications, emails, voicemails and faxes are being provided using just one platform. All these services can be accessed through ordinary telephones and even via payphones.

These technological advances have changed business operations and the way people communicate. They have introduced new efficiencies into old services and created numerous new services. Consequently, ICTs have assumed an important place in the development of businesses, societies and nations, and have strongly impacted on quality of life.

In this report, the term ICTs covers a wide group of technologies and services, including:

- all types of computer, telecommunications and related equipment production;
- all types of computer, telecommunications and related research and development (R&D);
- all types of software production;
- all types of computer, telecommunications and related technical support and maintenance;
- all types of telecommunications and teledata services, including transmission of voice, data, video etc;
- all types of telecommunications and teledata network maintenance, control and billing;
- all types of media online and offline services, including book publishing, magazines, newspapers etc; launching and maintaining websites, web portals etc;
- all types of online and offline advertisements.

ICTs are the most rapidly-growing and technologically-changing component of the world industry and service sector. In 2001, the total turnover of the ICT sector of OECD member countries reached almost 2.1 trillion US$, with telecommunications and other ICT services leading the sector (see, graph below).

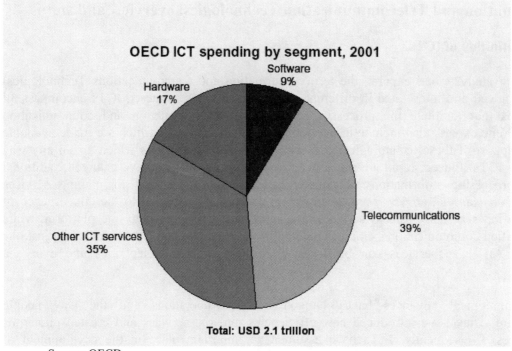

Source: OECD

1.2. ICT global trends

The prevailing trends and tendencies in the development of ICTs throughout the world are as follows:

- ICTs are beginning to play an increasingly important role in the world economy;
- ICTs account for a growing share of world investment, and their contribution to world output and productivity is growing rapidly;
- despite current cyclical difficulties, the growth of the ICT sector remains strong, and, according to OECD figures, the overall market for ICT goods and services continues to expand at an average annual rate of 8.3%;
- the ICT sector is the most globalised part of the world economy. Cross-border investments in this sector are on the increase, and ICT products represent more than one quarter of world imports and one-fifth of world exports;
- the centre of international investment in ICTs is shifting from manufacturing towards services;
- technology-oriented mergers, acquisitions and strategic alliances in the ICT production sector are happening as a result of rapid technological changes, a shortening of product life cycles and the opening of new markets;
- software is one of the most rapidly growing and evolving segments of the ICT sector. World software markets were estimated at US$ 196 billion in 2001;
- patenting relating to software is rapidly increasing. In the United States, for example, the number of software-related patents has grown much faster than any other patent area and now accounts for between 4% and 10% of the total number of patents;
- trade in software is dynamic but difficult to measure;

- electronic commerce has the potential to transform economic activity, but transactions are taking off more slowly than predicted. However, electronic transactions are on the rise. The Internet is increasingly used for purchasing. Total Internet sales in 2000 ranged between 0.4% and 1.8% of total sales;
- Internet transactions remain concentrated in a few areas of business operations, mainly in business-to-business. Business-to-consumer Internet sales remain low;
- the demand for ICT skills continues to grow. Governments, firms and non-profit organisations are taking measures to meet changing skill demands in ICT;
- differences in access to ICTs, such as in the availability and accessibility of computers and telecommunications services, create a disproportion and even a division between those able to benefit from opportunities provided by ICTs and those who cannot;

To better understand the above trends, both the production and the services sides of ICTs need to be considered:

The production side of ICT is purely technological and is represented by ICT goods. The services side of ICT, on the other hand, is represented by types of services, areas of application and categories of users (households, governments, businesses etc). The sections below highlight the major trends in these two sides of ICTs.

1.3. ICT technological trends

Over the last decade, the ICT sector has experienced major changes at a technological level. One of the most striking changes has been the upsurge in new technologies for telecommunications and related industries. Importantly, there has also been a persisting decline in the price of existing ICT technologies.

The main trends and tendencies in information and telecommunication technologies throughout the world are as follows:

- new generations of computers are appearing and computing potential is being developed. As computing power increases, unit price and size decreases and communication capabilities expand;

- a new generation of telecommunication standards and recommendations is appearing, including standards for mobile communications, networking, broadband access, digital broadcasting, control and billing;

- in all types of broadband access, a revolution is occurring in the "last broadband mile" to customer: digital subscriber line (DSL); broadband wireless access (BWA); cable access; satellite access etc.; are all involved in this revolution leading to a drop in prices for subscriber equipment and, subsequently, for services;

- new digital broadcast and media standards are appearing opening up new possibilities, such as interactive modes of operation, higher density, digital and Internet compatibility, miniaturisation etc;

- new generations of mobile technologies, such as 3G, are appearing. Notably, these can be integrated and are compatible with data and video;

- new strategic software, hardware and standards are appearing in the fields of control, management and billing.

All these changes in technologies lead to changes in services for all types of customers, from private customers and small/home office users (SOHO) to large corporations and State organisations.

1.4. ICT services trends

The revolution in ICT technologies has opened up new possibilities in telecommunications services at new levels of quality and price. Trends reflecting this are as follows:

- the price of Internet access is decreasing rapidly. Customers are increasingly being encouraged to consider channels that offer quality of service (QOS) and committed information rate (CIR) rather than the unit cost of Mbytes or Gbytes;

- mobile communications are now widely spread not only in highly developed countries, but also in countries whose economies are in transition. This includes data and e-document transmission services, such as SMS, WAP, GPRS, 3G-mobile, all of which are either quickly penetrating mobile services or are set to do so;

- with increasingly global computing power and telecommunications capacity, the dominant model of information exchange is shifting from a centralised and hierarchical model to a more decentralised, horizontal, equally distributed and democratic one. Open source, Internet protocol version 6 (IPv6), wireless and peer-to-peer are some examples of different aspects of the shift in the structure and nature of information exchange;

- there appears to be a transition from selling computers to customers to giving them away free of charge, or at a token cost, with monthly payments to access services;

- in highly-developed countries, digital and interactive TV offering completely new services for customers, are very popular and are penetrating the market rapidly;

- in highly-developed countries there is a strong trend towards creating clusters or zones of public access to Internet services with wireless technologies;

- electronic turnover of documents is developing not only in areas where it was first seen as natural, such as e-commerce, e-banking, e-advertising and e-government, but also in sectors where its uses were not originally anticipated, creating e-health, e-education and e-culture.

1.5. ICT measurements and statistics

On both the production and services side of ICT a large quantity of measurements and data are now available.

To assess the importance of ICT production overall, OECD and UNIDO ranked their member countries by the following indicators: employment; value added; R&D; and trade; giving each indicator an equal weight. Three categories of countries were created: High, Middle and Low ICT intensity.

As a result of statistical analysis, Finland, Hungary, Ireland, Korea, Sweden, UK, and USA were identified as belonging to the high intensity country grouping. Australia, Belgium, the Czech Republic, Germany, New Zealand, Poland, Portugal, Spain, and Turkey were identified as belonging to the low intensity country grouping.

UNIDO and OECD have also measured the importance of ICT manufacturing branches in total manufacturing. These measurements take into account both OECD countries and developing countries. The main findings are as follows:

- Singapore is outstanding in terms of a manufacturing specialisation in ICT industries. Almost 46% of total manufacturing performed in Singapore comes from ICT branches, primarily from the manufacturing of office, accounting and computing machinery (24%) and from electronic valves and tubes and other electronic components (16%). A specialisation of comparable strength is not seen in any other country, even within the OECD, where the most specialised countries are Korea, Finland, Ireland (16%) and Japan (14%);

- Asia has the greatest prevalence of developing countries with a degree of specialisation in ICTs. Thailand follows Singapore and Korea with an ICT share in total manufacturing of 38 %;

- Latin America's manufacturing owes very little to ICT branches;

- in Africa, ICTs are present only in few countries.

Some generalised OECD data on ICT production is shown in three graphs below.

The first graph shows ICT intensity in different OECD countries and in different segments of the ICT sector (telecommunications, other ICT services, software and hardware) for 2001 as a percentage share of total ICT in country's GDP.

The second graph shows the growth of trade in ICT goods from 1990 to 2000 looking at total trade in ICT as well as ICT major product groups (computer equipment, communications equipment and electronic components). All results are indexed with 1990 results equal to 100 percent.

The third graph shows the percentage of computer workers in some OECD countries and regions in 1995 and in 1999.

Official statistics assess the consumption side of ICT services. [Statistics Bureau and Statistics Center, 2001]. Three major types of consumers are identified: households, government and business sectors.

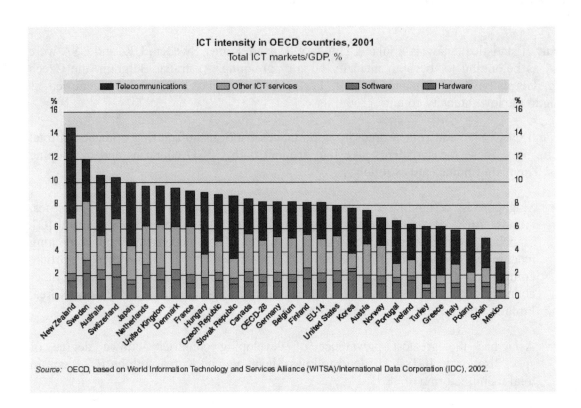

ICT intensity in OECD countries, 2001
Total ICT markets/GDP, %

Source: OECD, based on World Information Technology and Services Alliance (WITSA)/International Data Corporation (IDC), 2002.

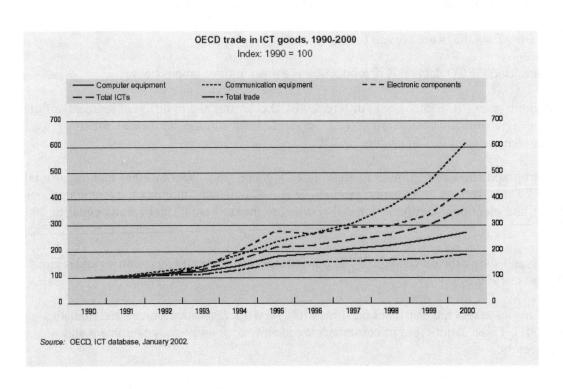

OECD trade in ICT goods, 1990-2000
Index: 1990 = 100

Source: OECD, ICT database, January 2002.

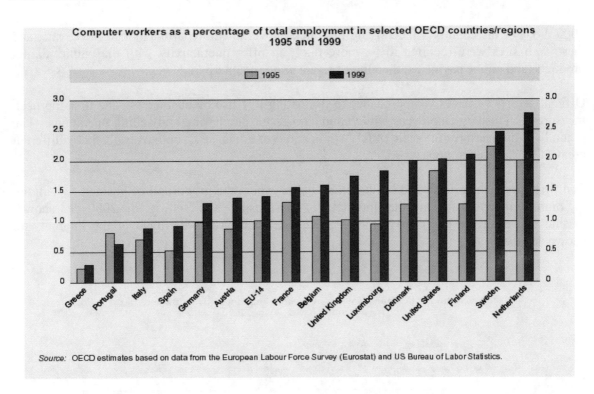

Computer workers as a percentage of total employment in selected OECD countries/regions 1995 and 1999

Source: OECD estimates based on data from the European Labour Force Survey (Eurostat) and US Bureau of Labor Statistics.

Several countries have already initiated data collection by adding questions concerning ICTs onto ongoing surveys or to censuses, or by conducting dedicated ICT surveys. These include questions concerning: the ownership of ICT facilities and the use of PC/Internet by households or individuals; purchases of telecommunications facilities, such as PCs or mobile phones and ongoing expenditures for their use, such as telecommunications charges; the place and time of PC and Internet use by households or individuals, and what purposes they serve for these sectors. The results of such data collection demonstrate strong trends:

- ICT use spreads extremely rapidly. In Korea, for example, computer ownership by households rose from 29% in 1997 to 46% in 1999. Computer literacy rates among people aged six years old and over soared from 40% in 1997 to 52% in 2000. The hours spent on computer per week increased from six in 1997 to 17 in 2000. In Japan, in 2000, 75% of households owned some ICT facility (mobile phone, word processor, fax, car navigation) and more than 50% had a PC. ICT expenditure per household was on average 3.7% of total expenditure. Internet users aged above 15 years increased from 12 million in 1997 to 47 million in 2000, thus attaining a penetration rate of 37%. In Canada, in 2000, 53% of individuals aged 15 years old or more used the Internet;

- the digital divide passes mostly across age and income lines. In Japan, for example, 70% of 15 to 19 years old, and 79% of 20 to 29 years old, use the Internet, but only 34% of 50 to 59 years old, and 15% of 60 to 69 years old are users.

At this time, few results are available on Government use of ICTs. In Australia, it was found that, by mid-1998, 89% of federal departments had a web site and that 12% received, and 33% placed, orders via the Internet. The Survey of Electronic Commerce and Technology of Canada came up with data suggesting that, in Canada, the government plays a lead role in Internet use. In

2000, 96% of the country's federal and provincial government departments had a web site, and these web sites were reported to be more likely to offer interactivity with the public, digital products, and online payments, than their private sector counterparts

Official statistics offers two approaches to evaluating ICT use by businesses. The first approach measures ICT use by employees while conducting their purchasing and selling operations. The second option measures the scale of ICT application (or e-business), both overall and by different areas of business operations.

The first approach is illustrated by the graph below where the proportion of businesses with ten or more employees, using the Internet for purchasing and selling in 2000, is shown. Businesses, receiving orders over the Internet, and businesses ordering over the Internet, are shown separately.

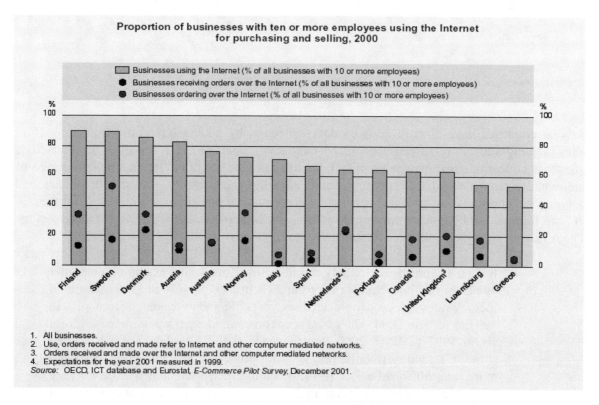

Proportion of businesses with ten or more employees using the Internet for purchasing and selling, 2000

☐ Businesses using the Internet (% of all businesses with 10 or more employees)
● Businesses receiving orders over the Internet (% of all businesses with 10 or more employees)
● Businesses ordering over the Internet (% of all businesses with 10 or more employees)

1. All businesses.
2. Use, orders received and made refer to Internet and other computer mediated networks.
3. Orders received and made over the Internet and other computer mediated networks.
4. Expectations for the year 2001 measured in 1999.
Source: OECD, ICT database and Eurostat, E-Commerce Pilot Survey, December 2001.

In the second approach e-business can be disaggregated into the following categories: e-manufacturing (how e-manufacturing of products is supported by computer networks); e-management (how business is conducted); and e-commerce (selling goods and services online). Innovation, too, is supported by ICT and could be identified as a particular process.

The spread of e-business can be evaluated by the total number of online shoppers, for example, or, business-to-business transaction volume, or by an indirect measurement such as the number of hosts per 1000 inhabitants (see graph below).

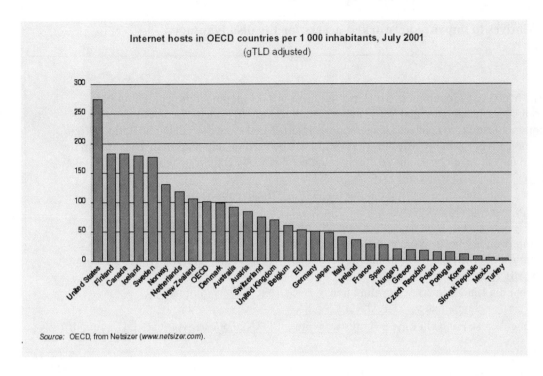

Internet hosts in OECD countries per 1 000 inhabitants, July 2001
(gTLD adjusted)

Source: OECD, from Netsizer (www.netsizer.com).

1.6. Success stories

ICT success stories and case studies provide an illustration of modern day challenges and their solutions in individual countries. They target a wide audience, ranging from regulators and corporate managers to academics. The studies help to set context, define the issues, analyse various models or strategies that can be employed, describe current approaches, identify obstacles and assess past and present experiences. The success stories of other regions of the world (not Eastern Europe and CIS) have been highlighted in this report in order to raise awareness among international organisations, government officials, regulators and industry players in developing countries, and especially in countries of Eastern Europe and CIS. They also provide policy-makers and other stakeholders with in-depth substantive information on how the problems of the digital divide are being solved in other parts of the world.

The following success stories are mainly taken from the ICT website www.itu.org which highlights some of the ongoing and successful ICT development projects taking place around the world.

1.6.1. Networked access

The most popular and efficient way for remote and poorly-connected clusters and communities to access the global information network is through public access points (PAPs). These PAPs include multipurpose community telecenters (MCTs), telecottages, and traditional cybercafés. These are often located in low-income areas of the world. With the aim of developing the human capacities of a given society, and of improving our understanding of how ICTs can be used for sustainable development, these PAPs employ a variety of ICTs (e.g. computers, photocopiers, Internet access) to bring the digital age to marginalised communities. Moreover, many PAPs also provide users with a variety of business services, including computer training, Web hosting and even distance education. Some of the projects, supported by ITU, are provided in Box 1 below.

Box 1. Initiatives to improve networked access for the disadvantaged.

- The Malaysian Government's e-Bario Initiative is a development project that utilises computers, telephones, and VSATs (very small aperture terminals for satellite communication) to connect villagers to the Internet in the remote village of Bario. E-Bario demonstrates the many ways in which ICTs can be used to help marginalised communities in Malaysia develop socially, culturally and economically (http://www.itu.int/ITU-D/ict/cs/malaysia/material/MYS CS.pdf.);
- The Leland Initiative to bring the Internet to the African continent (http://www.usaid.gov/leland.).
- MCT`s project in Uganda, (http://www.itu.int/ITU-D/ict/cs/uganda/material/uganda.pdf.).
- Empowering Mayan Women, Solola, Guatemala, (http://www.rockfound.org/display.asp?context=1&Collection=3&DocID=423&Preview=0&ARCurrent=1).
- Cambodia Khmer Internet Development Service, (http://www.itu.int/ITU-D/ict/cs/cambodia/material/KHM%20CS.pdf).
- Public Access Points in Egypt, (http://www.itu.int/ITU-D/ict/cs/egypt/material/egypt.pdf).

1.6.2. Networked learning

Given the limited educational resources to be found throughout the developing world and in transitional economies, ICTs present a unique opportunity to deliver educational training to marginalised societies efficiently and in a cost-effective manner. Worldwide, there are a variety of innovative, ICT-focused development initiatives that seek to modernise countries' educational systems and prepare students for participation in the global information society, particularly, at the primary and secondary school levels. The success stories highlighted below show how the international community is helping historically disconnected and remote societies cross the digital divide and join the digital age (see Box 2).

Box 2. E-learning and e-teaching initiatives.

- Inaugurated in 1926, the Radin Mas primary school has become a pillar of high-technology in Singapore's education system. The school, which serves as a model for educational institutions of the future, has some 200 computers, most of which are connected to the Internet via Singapore One's high-speed domestic backbone and ADSL lines. With a computer-to-student ratio of 1:5, Radin Mas is one of the best-connected primary schools in the world. Rather than simply using computers as a reference tool, the focus of Radin Mas is to deeply integrate computers into the learning and creative process. For example, the 2,000-plus students who attend the school are encouraged to engage in cross-cultural e-mail exchanges with students from around the world, create e-cards for mother's day, make music with midi-enabled keyboards, and use the two dozen iMacs to experiment with digital art. Some of the 9-11 year olds have even created a virtual zoo (http://www.itu.int/ITU-D/ict/cs/singapore/material/Singapore.pdf).
- Technology savvy youth Korea efforts to create newly wired generation (http://www.sunrint.hs.kr);
- Digital scouts in Indonesia (http://www.itu.int/ITU-D/ict/cs/indonesia/material/IDN CS.pdf).
- Computer literacy project Uganda connect (http://www.itu.int/ITU-

> D/ict/cs/uganda/material/uganda.pdf).
> - Radio education project, realised in Columbia
>
> - (http://www.rockfound.org/display.asp?context=1&Collection=3&DocID=423&Preview=0 &ARCurrent=1).
> - Updating Ethiopia's educational system (http://www.avu.org/);
> - Bringing the Internet to LAO PDR`s schools (http://www.itu.int/ITU-D/ict/cs/laos/material/LAO CS.pdf.);
> - Malaysia`s multimedia university (http://www.itu.int/ITU-D/ict/cs/malaysia/material/MYS CS.pdf.);
>
> - Smart schools in Malaysia (http://www.itu.int/ITU-D/ict/cs/malaysia/material/MYS CS.pdf.)

1.6.3. Networked society

Electronic health information and telemedicine have begun to play a critical role throughout the world, particularly in the case of remote villages in the developing world with poor access to hospitals and medical centres. The Internet, CD-ROMs, digital cameras, and electronic databases provide doctors and patients all over the world with the resources and information to improve the well-being of people living on the periphery of global society. Some ITU examples of electronic health information and telemedicine are as follows (see Box 3):

Box 3. E-health initiatives.

> - HealthNet (www.HealthNet.org.np) is an international NGO, providing health care information and communication services in the developing world. HealthNet Nepal was launched in 1994, and has since become a premier provider of health care-related information to over 500 users in 134 organisations throughout the kingdom. By providing services such as: e-mail, reports, computerised medical records and statistics, and a comprehensive library of medical information and useful links, HealthNet gives users access to a wealth of information on how to protect themselves from a variety of viruses and diseases. For instance, Nepal has a chronic problem with water contamination. HealthNet provides a tool for users to arm themselves with information on how to avoid or effectively address illnesses related to a given water problem. HealthNet is currently in the process of updating and expanding its services to all of the country's health care practitioners, (http://www.itu.int/ITU-D/ict/cs/nepal/material/nepal.pdf).
> - Telemedicine project in Cambodia (http://www.itu.int/ITU-D/ict/cs/cambodia/material/KHM%20CS.pdf).
> - Ciranet.com project in Egypt, created by Citibank and Raya Holdings, (http://www.ciranet.com and/or http://www.itu.int/ITU-D/ict/cs/egypt/material/egypt.pdf).

1.6.4. Networked economy

E-commerce is developing extremely rapidly both in developed and developing countries. In developed countries, e-commerce is a new instrument that complements the existing old instruments of commerce, while, in developing countries, it is an instrument to overcome cultural, infrastructure, regulatory, linguistic and economic barriers. E-commerce gives marginalised peoples an opportunity to market and sell their products on the global market, thus aiding human, social and economic development. Moreover, e-commerce also serves to empower women who often represent the least connected segment of society.

In developing countries, such as Cambodia, India and Ethiopia, entrepreneurs are utilising the Internet to market their handicrafts and become players in the global economy. In many such countries, where banks are often unable, or unwilling, to grant credit, people at the lower end of the income spectrum do not have easy access to credit. Through the Internet, these groups have also been able to establish lines of credit with banking institutions abroad and domestic micro-financing sources, thus, affording them an opportunity to "go global". (http://www.itu.int/ITU-D/ict/cs/cambodia/material/KHM%20CS.pdf.,http://www.k2crafts.com/., http://www.itu.int/osg/spu/casestudies/ETH%20CS1.pdf.)

Many similar examples can be seen on the ITU website (www.itu.org).

1.6.5. Networked Policy

Throughout the world, governments are using ICTs and specifically the Internet for developing public sector processes and providing citizens with easier access to government services. This improved information flow between the public and private sectors has helped to establish an environment of trust between citizens and elected officials in many parts of the world. Furthermore, many ministries and institutions responsible for ICTs and modernisation initiatives have begun to realise the importance of establishing a national strategy for becoming a participant in the emerging global information society. Some ITU examples of these new policies are highlighted below (see Box 4):

Box 4. ICTs for Democratisation.

- SMS, or short message service, enables mobile phone users to send short text messages to each other. Filipino users are responsible for about 10% of total global SMS traffic, or 50 million SMS messages per day, making Filipino society the largest society of SMS users in the world. In early 2001, SMS played a crucial role in the revolt called "People Power 2", whereby Filipinos used SMS to coordinate demonstrations that eventually led to the ousting of President Estrada. This highlighted one of the many ways in which ICTs can be used for democratisation. (http://www.itu.int/ITU-D/ict/cs/malaysia/material/MYS CS.pdf.).
- Singapore's e-Citizen project (http://www.itu.int/ITU-D/ict/cs/singapore/material/Singapore.pdf.).
- Modernizing Malaysia's Government, (http://www.itu.int/ITU-D/ict/cs/malaysia/material/MYS CS.pdf.).
- Regulatory reform and universal access in Brazil, (http://www.itu.int/newsarchive/wtdc2002/Lighting_the_Way.html).

---

2. Information and Telecommunication Technologies and relevant issues in Eastern European Countries and CIS

2.1. Common status

This section of the report considers 27 countries of the UNECE region. All of these countries have economies in transition, but these differ greatly in terms of overall economic situation, market size, ICT infrastructure, population and territory size etc. As has been mentioned previously, the UNECE countries in transition can be broken down into three subgroups, but the boundaries between the subgroups are not clear. The first subgroup includes countries of Eastern Europe with a comparatively high level of economic development, such as Hungary, the Czech Republic, Slovenia, Poland and some others. The third group includes a number of countries of the former USSR which are in the early stages of transition. The second group combines the Russian Federation, Ukraine and the Republic of Belarus, which face many problems in consolidating transition reforms, but which are still countries with extremely high potential, especially for the development of a knowledge-based economy. Some common parameters of all these countries are shown in table 1. The table also shows comparative figures for the USA. Where possible, a comparison with the USA will be shown in all other tables in this report.

As shown in chapter 1, the international market for ICTs is already enormous. It is now growing at a rate of 8% per year. One recent study which emphasises the importance of ICTs in the world economy despite their having existed for little more than 20 years, notes that Eastern Europe and the Russian Federation have lagged behind other regions in terms of absolute expenditure on ICTs. This was one of the major conclusions of "Digital Planet", the latest annual report released by the World Information Technology and Services Alliance (WITSA). The survey also reveals that, as regards total spending on ICTs, Eastern Europe and the Russian Federation are at the bottom of the scale, and the Eastern countries are behind not only Western Europe and North America, but also such regions as the Middle East/Africa, Latin America and Asia-Pacific. According to this report, however, Eastern Europe and the Russian Federation are now moving rapidly to catch up. When it comes to growth in ICT spending, the Eastern countries and Latin America, the regions with the smallest ICT bases, are currently outpacing by a rate of two-to-one Western Europe and North America, regions with mature ICT structures.

Many analysts(eg. Nikola Krastev, Radio Liberty, see bibliography) support the belief that there are now three distinct ICT development groups among the Eastern European countries. The Czech Republic, Estonia, Hungary, Poland, Slovakia, and Slovenia are in the first, most advanced group. The second-ranking ICT regional group is composed of Bulgaria, Croatia, Latvia, Lithuania, Macedonia, Romania, and, to a lesser degree, Yugoslavia. The Russian Federation and Ukraine make up the third, least-developed group, each with special ICT circumstances. The UNECE report challenges the findings of the WITSA report, arguing that now the second group in the WITSA division has simply disappeared. Some of its countries, such as Latvia, have moved up to the first group, and some have approximately the same ICT parameters as the countries of the third group in the WITSA division. It also argues that some Eastern European countries and, to some degree, the Russian Federation, have had some success in catching up with the developed countries

Table 1. Population and GDP: Eastern European Countries, CIS and USA, 2001.

No	Country	Population (millions) total 2001	Population density per sq. km 2001	GDP total (Billions, US$) 2000	GDP per capita (US$) 2000
1.	Armenia	3.79	126	1.9	544
2.	Azerbaijan	7.78	90	4.0	514
3.	Georgia	5.47	78	2.9**	526
4.	Kazakhstan	16.09	6	15.8**	973
5.	Kyrgyz Republic	4.99	25	1.2**	255
6.	Tajikistan	6.13	43	1.1**	178
7.	Turkmenistan	4.84	10	2.5*	582
8.	Uzbekistan	25.26	56	15.4**	676
9.	Albania	3.97	138	3.7	940
10.	Republic of Belarus	10.25	49	8.3	814
11.	Bosnia	4.07	80	4.5**	1178
12.	Bulgaria	8.11	73	12.0	1473
13.	Croatia	4.66	82	19.0	4253
14.	Czech Republic	10.27	130	50.8	4931
15.	Estonia	1.43	32	5.0	3455
16.	Hungary	9.97	107	45.6	4561
17.	Latvia	2.35	37	7.1	2930
18.	Lithuania	3.68	56	11.2	3042
19.	Moldova	4.39	130	1.3	294
20.	Poland	38.63	124	157.6	4078
21.	Romania	22.39	94	36.7	1636
22.	Russian Federation	146.76	9	251.1	1709
23.	Slovak Republic	5.40	110	19.1	3540
24.	Slovenia	2.00	99	18.1	9108
25.	TFYR Macedonia	2.04	79	3.4**	1705
26.	Ukraine	50.30	83	30.8**	608
27.	Yugoslavia	10.68	105	11.3*	1067
28.	United States	285.93	31	9962.6	36211

*1998 **1999

2.2. Measurement and statistics: the whole region

There are different ways of measuring the level of ICT development in the UNECE region and in that of the countries in transition. This section tries to evaluate some of the most important parameters of ICT development, using ITU and World Bank data, transforming and consolidating it without changing the raw numbers. As in the previous chapter, the aim is to measure both the production and services sides of ICT.

To gain an overall impression of ICT production, the World Bank has tried to evaluate the most important parameters of such production in the UNECE member countries and in the countries in transition using the following indicators: (i) total ICT production, (ii) ICT percentage of GDP, (iii) ICT per capita, (iv) scientists and engineers in R&D, (v) expenditures on R&D. It is important to note that the United Nations Industrial Development Organisation (UNIDO) has also attempted to measure the level of ICT development by ranking countries into three categories of ICT intensity - High, Medium and Low. The same categories, as was previously suggested, may be applied to the three subgroups of the UNECE countries in transition. The results of such a statistical manipulation are presented in Table 2. There is a problem, however, with the availability of relevant data. Many countries are not measured.

More data is available for measuring the development of the ICT services side for the UNECE countries in transition. In tables 3 to 5 the following indicators are presented: the level of development of Internet access; main telephone networks; and mobile networks in the transition countries. All tables use ITU statistics and clearly show the existing situation. In some cases, it was also possible to evaluate growth within a five to six year period. The measurements include all three consumer groups within for ICT services – households, Government and the business sector.

Table 2. ICT total - World Bank statistics, 2001

No	Country	Total ICT (US$, millions)	ICT as % of GDP	ICT per capita	Scientists and engineers in R&D per mill. of people	Expenditures on R&D (% of GNP)
1.	Armenia	N/A	N/A	N/A	1307.8	0.2
2.	Azerbaijan	N/A	N/A	N/A	2735.3	N/A
3.	Georgia	N/A	N/A	N/A	N/A	N/A
4.	Kazakhstan	N/A	N/A	N/A	N/A	0.3
5.	Kyrgyz Republic	N/A	N/A	N/A	573.6	0.2
6.	Tajikistan	N/A	N/A	N/A	N/A	N/A
7.	Turkmenistan	N/A	N/A	N/A	N/A	N/A
8.	Uzbekistan	N/A	N/A	N/A	N/A	N/A
9.	Albania	N/A	N/A	N/A	N/A	N/A
10.	Republic of Belarus	N/A	N/A	N/A	2295.7	0.6
11.	Bosnia	N/A	N/A	N/A	N/A	N/A
12.	Bulgaria	530.0	3.8	65.4	1289.0	N/A
13.	Croatia	N/A	N/A	N/A	1494.0	N/A
14.	Czech Republic	4954.0	9.5	483.4	1319.9	1.3
15.	Estonia	N/A	N/A	N/A	2164.2	0,8
16.	Hungary	4346.0	8.9	465.5	1249.4	0.7
17.	Latvia	N/A	N/A	N/A	361.5	N/A
18.	Lithuania	N/A	N/A	N/A	2030.7	N/A
19.	Moldova	N/A	N/A	N/A	334.4	0.8
20.	Poland	10489.0	5.9	271.1	1460.0	0.7
21.	Romania	956.0	2.2	42.8	N/A	N/A
22.	Russian Federation	9908.0	3.3	68.2	3397.1	1.1
23.	Slovak Republic	702.0	4.0	130.8	1706.0	N/A
24.	Slovenia	547.0	2.9	275.4	2161.3	1.5
25.	TFYR Macedonia	N/A	N/A	N/A	387.2	0.3
26.	Ukraine	N/A	N/A	N/A	2120.6	1.0
27.	Yugoslavia	N/A	N/A	N/A	2389.3	1.3
28.	United States	812,635.0	7.9	2923	4102.9	2.5

Table 3. Internet access - ITU statistics, 2001

No	Country	Internet hosts total	Internet hosts per 10,000 inhabit.	Internet users (000) total	Internet users per 10,000 inhabit.	Estimated PCs (000) total	Estimated PCs per 100 inhabit.
1.	Armenia	2361	6.23	50.0	142.05	30.0	0.79
2.	Azerbaijan	1314	1.69	25.0	32.13	N/A	N/A
3.	Georgia	2081	3.80	25.0	45.7	N/A	N/A
4.	Kazakhstan	10947	6.80	100.0	61.64	N/A	N/A
5.	Kyrgyz Republic	4558	9.14	51.6	105.74	N/A	N/A
6.	Tajikistan	299	0.49	3.2	5.22	N/A	N/A
7.	Turkmenistan	1620	3.35	8.0	16.55	N/A	N/A
8.	Uzbekistan	213	0.08	150.0	59.39	N/A	N/A
9.	Albania	187	0.47	10.0	25.18	30.0	0.76
10.	Republic of Belarus	3287	3.21	422.2	411.87	N/A	N/A
11.	Bosnia	3248	799	450	40.65	N/A	N/A
12.	Bulgaria	26926	33.21	605.0	746.27	361*	4.43
13.	Croatia	21988	47.24	250.0*	558.91	400	8.59
14.	Czech Republic	215525	209.78	1400.0	1362.66	1250*	12.14
15.	Estonia	51040	356.92	429.7	3004.59	250	17.48
16.	Hungary	167585	168.04	1480.0	1484.01	1000	10.03
17.	Latvia	25003	106.35	170.0	723.10	360	15.31
18.	Lithuania	35155	35.50	250.0	679.16	260	7.06
19.	Moldova	1756	4.00	60.00	136.67	70	1.59
20.	Poland	489895	126.82	3800.0	983.72	3300	8.54
21.	Romania	46283	20.67	1000.0	446.63	800	3.57
22.	Russian Federation	354339	24.14	4300.0	293.3	7300	4.97
23.	Slovak Republic	72557	134.29	650.0*	1203.26	800	14.81
24.	Slovenia	29550	148.16	600.0	3007.52	550	27.57
25.	TFYR Macedonia	2594	12.69	70.0	342.47	N/A	N/A
26.	Ukraine	58235	11.58	600.0	119.29	920	1.83
27.	Yugoslavia	15664	14.67	600.0	561.8	250	2.34
28.	United States	106,193,339	3714.01	142,823.0	4995.10	178,000	62.25

*2000

Table 4. Main telephone lines - ITU statistics, 2001

No	Country	1995 total in thousands	2001 total in thousands	CAGR (%) 1995-2001 total	1995 per 100 inhabit.	2001 per 100 inhabit.	CAGR (%) 1995-2001 per 100 inhabit. .
1.	Armenia	582.8	529.3	-1.6	15.45	11.19	-4.6
2.	Azerbaijan	639.5	865.5	5.2	8.49	11.13	4.6
3.	Georgia	554.3	867.6	7.8	10.23	15.86	7.6
4.	Kazakhstan	1962.9	1834.2*	-1.3	11.87	11.31	-1.0
5.	Kyrgyz Republic	357.0	376.1*	1.0	7.92	7.71*	-0.5
6.	Tajikistan	262.7	223.0	-2.7	4.50	3.63	-3.5
7.	Turkmenistan	320.3	387.6	3.2	7.14	8.02	1.9
8.	Uzbekistan	1544.2	1663.0	1.2	6.81	6.58	-0.6
9.	Albania	42.1	197.5	29.4	1.17	4.97	27.3
10.	Republic of Belarus	1968.4	2857.9	6.4	19.18	27.88	6.4
11.	Bosnia	237.8	450.1	11.2	5.99	11.07	10.8
12.	Bulgaria	2562.9	2913.9	2.2	30.47	35.94	2.8
13.	Croatia	1287.1	1700.1	4.7	28.28	36.52	4.4
14.	Czech Republic	2444.2	3846.0	7.8	23.65	37.43	9.0
15.	Estonia	411.7	503.6	3.4	27.74	35.21	4.1
16.	Hungary	2157.2	3730.0	9.6	21.05	37.4	10.1
17.	Latvia	704.5	724.8	0.5	27.65	30.83	1.7
18.	Lithuania	941.0	1151.7	3.4	25.35	31.29	3.6
19.	Moldova	566.5	676.1	3.0	13.02	15.40	7.8
20.	Poland	5728.5	11400.9	12.2	14.84	25.51	12.1
21.	Romania	2968.0	4094.0	5.5	13.09	18.28	5.7
22.	Russian Federation	25018.9	35700.0	6.1	16.91	24.33	6.3
23.	Slovak Republic	1118.5	1556.3	5.7	20.84	28.80	5.5
24.	Slovenia	614.8	799.7	4.5	30.93	40.09	4.4
25.	TFYR Macedonia	351.0	538.5	7.4	17.85	26.35	6.7
26.	Ukraine	8311.0	10699.6	4.3	16.09	21.21	4.7
27.	Yugoslavia	2017.1	2443.9	3.3	19.15	22.88	3.0
28.	United States	159,735.2	190,000.0	2.9	60.73	66.45	6.4

*2000

Table 5. Mobile cellular subscribers - ITU statistics, 2001

No	Country	1995 total in thousands	2001 total in thousands	CAGR (%) 1995-2001 total	2001 in thousands per 100 inhabit.	2001 % of digital	2001 % of total telephone subscribers
1.	Armenia	N/A	25.0	N/A	0.66	100.0	4.5
2.	Azerbaijan	6.0	620.0	116.6	7.97	N/A	41.7
3.	Georgia	0.1	295.0	254.0	5.39	N/A	25.4
4.	Kazakhstan	4.6	582.0	124.1	3.62	N/A	25.4
5.	Kyrgyz Republic	N/A	27.0	N/A	0.54	N/A	24.1
6.	Tajikistan	N/A	1.6	N/A	0.03	N/A	0.7
7.	Turkmenistan	N/A	9.5*	N/A	0.21	N/A	2.5
8.	Uzbekistan	3.7	62.8	60.1	0.25	N/A	3.6
9.	Albania	N/A	350.0	N/A	8.82	100.0	63.9
10.	Republic of Belarus	5.9	138.3	69.2	1.35	85.7	4.6
11.	Bosnia	N/A	233.3	N/A	5.74	100.0	34.1
12.	Bulgaria	20.9	1550.0	104.9	19.12	N/A	34.7
13.	Croatia	33.7	1755.0	93.3	37.70	N/A	50.8
14.	Czech Republic	48.9	6769.0	127.4	65.88	N/A	63.8
15.	Estonia	30.5	651.2	66.6	45.54	N/A	56.4
16.	Hungary	265.0	4968.0	63.0	49.81	N/A	57.1
17.	Latvia	15.0	656.8	87.7	27.94	N/A	47.5
18.	Lithuania	14.8	932.0	99.5	25.32	N/A	44.7
19.	Moldova	N/A	210.0	396.6	4.78	N/A	23.7
20.	Poland	75.0	10050.0	126.2	26.02	N/A	46.9
21.	Romania	9.1	3860.0	74.3	17.24	N/A	48.5
22.	Russian Federation	88.5	5560.0	99.4	3.79	82.0	13.5
23.	Slovak Republic	12.3	2147.0	136.4	39.74	N/A	58.0
24.	Slovenia	27.3	1515.0	95.3	75.98	N/A	65.5
25.	TFYR Macedonia	N/A	223.3	N/A	10.92	100	29.3
26.	Ukraine	14.0	2224.6	132.7	4.42	N/A	17.9
27.	Yugoslavia	N/A	1997.8	N/A	18.71	N/A	45.0
28.	United States	33,785.7	127,000.0	24.7	44.42	N/A	40.1

*2000

ICT business and government environment indicators taken from the World Bank data statistics are shown in table 6. The World Bank ranked countries according to their level of ICT services development in business and government, ranging from 1 (the lowest) to 7 (the highest). This included some important parameters, such as: Internet speed and access, Internet effects on business, rapidly growing IT job market, competition in ISPs, government online services availability and laws related to ICT use. It is important to underline that these statistics are only available for comparatively developed countries.

Table 6. ICT Business and government environment - World Bank statistics, 2000

No	Country	Internet speed and access	Internet effects on business	High-speed IT job market	Competition in ISPs	Government online services availability	Laws relating to ICT use
1.	Armenia	N/A	N/A	N/A	N/A	N/A	N/A
2.	Azerbaijan	N/A	N/A	N/A	N/A	N/A	N/A
3.	Georgia	N/A	N/A	N/A	N/A	N/A	N/A
4.	Kazakhstan	N/A	N/A	N/A	N/A	N/A	N/A
5.	Kyrgyz Republic	N/A	N/A	N/A	N/A	N/A	N/A
6.	Tajikistan	N/A	N/A	N/A	N/A	N/A	N/A
7.	Turkmenistan	N/A	N/A	N/A	N/A	N/A	N/A
8.	Uzbekistan	N/A	N/A	N/A	N/A	N/A	N/A
9.	Albania	N/A	N/A	N/A	N/A	N/A	N/A
10.	Republic of Belarus	N/A	N/A	N/A	N/A	N/A	N/A
11.	Bosnia	N7A	N/A	N/A	N/A	N/A	N/A
12.	Bulgaria	3.5	3.1	2.5	4.0	3.0	3.0
13.	Croatia	N/A	N/A	N/A	N/A	N/A	N/A
14.	Czech Republic	4.1	4.5	5.2	6.0	5.8	4.8
15.	Estonia	6.0	4.5	5.2	6.0	5.8	4.8
16.	Hungary	4.0	3.3	4.8	5.3	4.4	3.8
17.	Latvia	3.6	3.5	4.1	4.8	3.3	3.0
18.	Lithuania	3.3	3.2	3.6	4.1	4.5	3.4
19.	Moldova	N/A	N/A	N/A	N/A	N/A	N/A
20.	Poland	2.9	3.7	5.3	4.7	3.1	3.9
21.	Romania	4.1	3.6	2.2	3.6	1.2	1.9
22.	Russian Federation	3.2	2.8	3.7	3.7	2.6	2.4
23.	Slovak Republic	4.4	3.6	4.8	5.2	3.4	3.2
24.	Slovenia	4.3	3.8	4.9	4.4	3.7	4.2
25.	TFYR Macedonia	N/A	N/A	N/A	N/A	N/A	N/A
26.	Ukraine	3.2	4.1	2.7	4.3	3.5	4.0
27.	Yugoslavia	N/A	N/A	N/A	N/A	N/A	N/A
28.	**United States**	**6.6**	**5.0**	**5.7**	**6.7**	**5.4**	**5.6**

World Bank statistics on the media and education environment in the countries in transition are considered vital to this report. These parameters are shown in table 7, and include daily newspapers, radio and television sets per 1000 people, networked PCs and PCs installed for education.

Table 7. Media and education - World Bank statistics, 2000/2001

No	Country	Daily newspapers per 1000 people	Radios sets per 1000 people	Television sets per 1000 people	Networked PCs (%)	PCs, installed for education (thousands)
1.	Armenia	23	221	216	N/A	N/A
2.	Azerbaijan	N/A	22	259	N/A	N/A
3.	Georgia	N/A	556	474	N/A	N/A
4.	Kazakhstan	N/A	422	241	N/A	N/A
5.	Kyrgyz Republic	15	111	49	N/A	N/A
6.	Tajikistan	20	141	326	N/A	N/A
7.	Turkmenistan	N/A	256	196	N/A	N/A
8.	Uzbekistan	3	456	276	N/A	N/A
9.	Albania	35	243	123	N/A	N/A
10.	Republic of Belarus	155	299	342	N/A	N/A
11.	Bosnia	N/A	243	111	N/A	N/A
12.	Bulgaria	257	543	449	22.1	44.2
13.	Croatia	114	340	293	N/A	N/A
14.	Czech Republic	254	803	508	99.6	35.5
15.	Estonia	176	1096	591	N/A	N/A
16.	Hungary	46	684	437	76.7	40.7
17.	Latvia	15	273	137	N/A	N/A
18.	Lithuania	30	524	422	N/A	N/A
19.	Moldova	154	758	297	N/A	N/A
20.	Poland	108	524	400	41.9	252.7
21.	Romania	N/A	334	381	36.8	31.1
22.	Russian Federation	105	418	421	51.0	471.3
23.	Slovak Republic	196	567	380	13.4	27.7
24.	Slovenia	196	372	327	13.3	28.8
25.	TFYR Macedonia	27	201	183	N/A	N/A
26.	Ukraine	101	889	456	N/A	N/A
27.	Yugoslavia	107	297	282	22.6	N/A
28.	**United States**	**213**	**2118**	**854**	**80.4**	**16322.7**

All of the above indicators are vital in order to evaluate the ICT situation and what progress is being made towards a knowledge-based economy in the region, and for understanding the primary development trends. However, the picture is incomplete due to the fact that some parameters, especially of countries in the third subgroup in transition, are not measured. Consequently, this report attempts to gain a complete picture using the data provided in the series of previously-mentioned country assessment reports "Towards A Knowledge-Based Economy".

There are only nine country reports available from the UNECE's 27 economies in transition. Importantly, however, the sample of the countries is quite representative.

2.3. Measurements and statistics derived from the country assessment reports "Towards A Knowledge-Based Economy"

All the country assessment reports evaluated the readiness of countries in transition for a knowledge-based economy by using 19 basic indicators combined in five major groups. While this report has employed the same approach and measurements, some further indicators were also used. In order to make a sound comparison of countries with very small populations, such as Latvia or Armenia, with those with large populations, such as the Russian Federation, the absolute figures from the country assessment reports were, in many cases, recalculated using the population weight. In addition, all statistics on sectors relevant to the development of the knowledge-based economy, such as education, intellectual capital, labour force etc. were used, as well as measurements and indicators on the ICT sector itself. The summary of all statistical materials gathered from the country assessment reports is presented below. Tables 8 and 9 combine the major indicators of knowledge-based technology for networked access in the countries in transition.

Table 8. Networked Access /information infrastructure

No	Country	Telephone mainlines per 100 people	Mobile wireless per 1000 people	Mobile wireless total	Cable TV subscribers total	Cable TV subscribers % of households
1.	Armenia	17.6	8.33	40000	N/A	N/A
2.	Republic of Belarus	30	10	235000	910000	30
3.	Bulgaria	39	226	1800000	1419148	48
4.	Georgia	13.4	78	355000	278000	25
5.	Kyrgyz Republic	7.8	4	17000	13753	12.9
6.	Latvia	31	342	800000	250000	31.25
7.	Russian Federation	21	68	11000000	12000000	20
8.	Slovak Republic	32	470	2600000	660000	35
9.	Yugoslavia	32.92	318.474	3000000	230000	2.72

Table 9. Networked Access /internet & software

No	Country	Total ISPs	Average Internet price per hour, (US$)	Average salary per month (US$)	Rates for 64 Kbps leased lines, (US$)	Local software (%)
1.	Armenia	17	0.6	N/A	120 - 180	5
2.	Republic of Belarus	47	0.9	100	540	5
3.	Bulgaria	>100	0.4	N/A	400	<20
4.	Georgia	8	0.5	36.6	500	0.2
5.	Kyrgyz Republic	12	0.6	N/A	500 for 32 Kbps	3
6.	Latvia	50	0.8	N/A	120 – 16 Kbps 740 – 2 Mbps	N/A
7.	Russian Federation	300	0.5	300	300	5 - 10
8.	Slovak Republic	5 large 25 small	US$2 per month	N/A	200	Low, 75% of virus protection
9.	Yugoslavia	48	0.5	150	520	N/A, growing trend

Tables 10 and 11 below show the results of the country assessment reports for four other major groups of indicators, such as networked learning, society, economy and policy. These tables demonstrate that the basic indicators were considered by all the national experts involved in the reports to be a viable means of evaluating and implementing e-strategies.

The rapid changes inherent in ICT make it necessary to analyse the status of the media, intellectual capital, the labour force, education and R&D in the countries in transition. Statistics and measurements for all of these fields except the media, are presented in tables 12 to 14. Clear statistics on the media in transition countries are available on the World Bank Website.

Table 10. Networked Learning

No	Country	Total no. of schools	Total no. of students	Average no. of PCs per school	Average no. of students per PC	Average no. of schools with PC labs (%)
1.	Armenia	1446	564000	0.225	1700	11
2.	Republic of Belarus	4574	1276000	10	41	74
3.	Bulgaria	1023	N/A	12	66	33
4.	Georgia	3467	749100	0.3	707	6.6
5.	Kyrgyz Republic	1980	1025792	1.6	330	15.556
6.	Latvia	1029	336824	13.3	25	91
7.	Russian Federation	68000	19190200	0.6	4800	20
8.	Slovak Republic	2302/1122/20	600885/ 289118/ 93971	3.13/ 10.31/ 304.3	25	22/317100
9.	Yugoslavia	4600	1043000	N/A, low	N/A, high	N/A, low

For the Slovak Republic, parameters are given as "primary/secondary/high".

Table 11. Networked Society/Economy/Policy

No	Country	% of people, using Internet regularly	Local domains per 1000 people	% of business and govern. Offices using PCs	No. of Websites of local business	No. of Govern. domains online	Status of liberalisation of ICT sector
1.	Armenia	2	0.1	N/A	94	<10	Not liberalised
2.	Republic of Belarus	3	0.2	50	2000	55	Liberalisation is planned
3.	Bulgaria	8.5	0.5	30	1860	150	Liberalisation is planned in 2003
4.	Georgia	5	0.6	15	N/A	About 10	Implementation stage
5.	Kyrgyz Republic	9.6	0.2	17.4	200	57.7% of agencies	Liberalisation is planned
6.	Latvia	21	3.38	135`000 PCs at workplace	7.4% of companies	Most of agencies	Liberalisation is planned in 2003
7.	Russian	3	0.5	90	2500	70	Implementation

						stage
8.	Slovak Republic	15.4	13	90	90% of companies	24	Implementation stage
9.	Yugoslavia	5.88	1.7	N/A	10000	Most of agencies	Liberalisation is planned

Table 12. Intellectual capital

No	Country	No. of patents per annum	No. of copyrights per annum	No. of licenses per annum	No. of trademarks per annum	No. of scientific and technical associations*
1.	Armenia	200	0	0	954	3
2.	Republic of Belarus	903	580	208	5155	105
3.	Bulgaria	N/A	N/A	205	N/A	N/A
4.	Georgia	150	0	15	100	5
5.	Kyrgyz Republic	16	996	439	295	7
6.	Latvia	323	N/A	47	7356	5
7.	Russian Federation	5500	2000	1200	800	9
8.	Slovak Republic	N/A	N/A	N/A	N/A	N/A
9.	Yugoslavia	853	0	N/A	4743	550

The large difference in the number of scientific and technical associations quoted may result from some national experts including only state organisations in this list while others have included all associations regardless of whether they are government bodies or not.

Table 13. Education/research and development

No	Country	No. of higher education establish-ments (public/ private)	Students per annum, total (public/ private)	Students In ICTs, (%)	Citizens with higher education (% of pop.)	No. of research institutes	Invest-ment in R&D per annum, (mill. USD)
1.	Armenia	54 (19/35)	60000 (43000/ 17000)	11	0.17 (science)	88	2.62
2.	Republic of Belarus	57 (43/14)	281.7 (36.6/245.1)	2.5	7.43	307	67
3.	Bulgaria	48	N/A	N/A	2.29	N/A	70
4.	Georgia	171 (26/145)	147000 (115000/ 31500)	7	17.9	62	4519
5.	Kyrgyz Republic	1980	1025792	0.351	N/A	90	N/A
6.	Latvia	34 (20/14)	110500 (88239/ 22261)	10.4	6.31	273	41.3
7.	Russian Federation	965 (607/358)	N/A	7	N/A	800	N/A
8.	Slovak Republic	24	About 18000	30.25	186000	15	70
9.	Yugoslavia	0.1	291	N/A 0.01	0.2	250	359

Table 14. Labour force

No	Country	% of emp. in science and tech.	Average salary per month, (US$)	% of emp. in educ.	Average salary per month, (US$)	% of emp. in ICTs	Average salary per month, (US$)
1.	Armenia	0.2	32	N/A	61	0.5	88
2.	Republic of Belarus	0.41	38.8	0.35	37.2	0.69	35.6
3.	Bulgaria	N/A	N/A	N/A	N/A	N/A	N/A
4.	Georgia	0.14	28	0.04	24	N/A	N/A
5.	Kyrgyz Republic	0.3	Average for country	N/A	N/A	0.02	295% of the national average
6.	Latvia	0.12	700	0.22	700	1000	0.28
7.	Russian Federation	N/A	N/A	N/A	N/A	0.3	N/A
8.	Slovak Republic	N/A	N/A	N/A	N/A	0.2	165% of the national average
9.	Yugoslavia	87	74500 (73500/ 1500)	28	3.74	212	6.4

Using the Harvard University benchmarking system, some national experts attempted to evaluate the level of knowledge economy achieved by their respective countries themselves. This was done by ranking all 19 indicators from 1 (the lowest stage) to 4 (the highest stage). In a few cases, some further parameters have been added by UNECE. The results of the ranking for all nine countries are laid out in table 15.

Table 15. Benchmarking

No	CATEGORY	Arm.	Bel.	Bul.	Georg.	Kyrgyz Rep.	Latvia	Russia	Slovak Rep.	Yugo slavia
	Networked Access	STAGE	STAGE	STAGE	STAGE	STAGE	STAGE	STAGE	STAGE	STAGE
1.	Information infrastructure	3	3	3	3	2	4/mob. 3/fixed	3	3	4/mobile 3/fixed 2/cable
2.	Internet availability	3	2	2	2	3	3	3	3	2/leased lines 3/other
3.	Internet affordability	3	2	3	2	2	3	3	3	2/leased lines 3/other
4.	Network speed and quality	3	2	3	2	3	3	3	3	3
5.	Hardware and software	3	3	3	2	2	3	3	3	3
6.	Service and support	3	2	4	2	3	3	3	3	3
	Networked learning									
7.	Schools´ access to ICTs	2	3	2/3	2	2	3	2	2/3	3/high 2/prim., second.
8.	Enhancing education with ICTs	2	3	2	2	2	3	2	2/3	3/high 2/prim., second.
9.	Developing the ICT workforce	2	2	3	2	3	3	2	2	3
	Networked society									
10.	People and organisations online	3	2	3	2	2	3/4	2	3	¾
11.	Locally relevant content	3	2	3	2	3	3	3	3	3
12.	ICTs in everyday life	3	2	3	3	3	3/4	3	2/3	3
13.	ICTs in the workplace	3	2	3	2	3	3	3	3	3
	Networked economy									
14.	ICT employment opportunities	2	2	3	2	2	3	3	2/3	2
15.	B2C Electronic Commerce	2	1	3	2	3	3	3	2	3
16.	B2B Electronic Commerce	2	1	3	2	3	3	3	2	2
17	E-government	2	2	3	2	2	3	3	2/33	2/3
	Networked policy									
18	Telecom. regulation	1	2	3	2	3	2	3	2/3	2
19	ICT trade policy	1	2	4	2	2	2	3	2/3	3

2.4. Global Trends

The above indicators and data, together with the findings of the nine country studies, allow certain conclusions to be reached on the current status and development of ICTs, and of related segments of economy and society, in the countries of Eastern Europe and CIS:

- all countries in transition can be broken down into three subgroups according to their level of development in the context of the knowledge-based economy – high, middle or low;

- in all countries in transition, ICTs and related segments of the economy are growing faster than others. Growth pace, however, is closely correlated with the level of economic development. This can be seen in tables 1 and 2. In the Czech. Republic, for example, the comparative GDP share of the ICT sector is 9.5% with GDP per capita standing at US$ 4931, while, in Romania, the GDP share of the ICT sector is only 2.2% and GDP per capita is US$1636;

- the countries of the first, or the highest group, are moving rapidly towards a knowledge-based economy, and the digital divide between them and the developed countries is not wide. There is a large divide, however, between these first-group countries and the rest of the economies in transition;

- The countries of the third, or the lowest subgroup are at the very beginning of their path towards a knowledge-based economy;

- the countries of the second or middle subgroup have enormous potential for building knowledge-based economies, but realisation of this potential is constrained by many internal and external factors;

- the role of the ICT sector and related (or interconnected) sectors is strong in all the countries in transition, but it is more clearly determined by the level of the knowledge-based economy achieved to date - the higher the level achieved, the bigger the role played by the ICT and related sectors;

- in the majority of the economies in transition, the ICT and interconnected sectors are more globalised than their counterparts in other parts of the world. This is especially true of ICT services. Many operators, carriers and ISPs have foreign investment, especially in mobile communications;

- as previously mentioned, the development of ICTs is shifting rapidly from production to services;

- the share of locally produced software is no more than 5 to 10% of the total software used in all countries of Eastern Europe and CIS. The rest is imported;

- the level of development of the ICT sector is closely connected with the liberalisation level of the ICT economy – the more liberalised the sector is, the higher is its level of development. All countries in transition are either in the process of liberalisation or at least are planning a liberalisation of their ICT sector in very near future;

- there are a number of reasons for the high demand for ICT specialists. These include the demand generated by the growth of the ICT production sector; the rapid growth in mobile communications; increased Internet and PC penetration; and the transition of enterprises to electronic accounting, etc. All these factors mean that ICT specialisation in high school is now one of the most interesting and prestigious career choices for youth. Initiatives have been launched to provide new interdisciplinary education and research support for businesses that are starting to, or already operating on, the emerging digitalised markets. The aim of these initiatives is to foster skills for emerging interactive applications in areas such as banking, insurance, telecoms, and media, but also for traditional manufacturing industries, where multimedia applications like sales promotion, will become a major tool. International education to satisfy the demands of the ICT sector is also developing both by networking and via more formalised cooperation with a number of universities in Europe and the United States, with whom some countries in transition, (the Russian Federation, Latvia, Hungary and others) now have common research projects and lecturer exchange programmes;

- in practically all the countries of Eastern Europe and CIS, structural reforms are in the implementation process, but are at different stages. The State, therefore, continues to play a more profound role in the economy, than in countries in other parts of the world. The State formulates informatisation policies, develops economic and legal mechanisms, and determines the rules of the game for everyone involved in the development of a knowledge-based economy. In the opinion of this report, the main regulation problem facing all countries in transition in developing a knowledge-based economy is not the degree of liberalisation which has been achieved, but rather the way the Government of a country has supported activities which lead to a knowledge-based economy, and how successful this support has been. The example of Armenia and Latvia illustrates this issue. For a number of years, Armenian Telecom (Armentel) has been controlled not by the Government, but by a foreign company, OTI, while Latvian Telecom (Lattelecon) has been controlled by the State. However, the situation with respect to activities leading to the formation of a knowledge-based economy is much better in Latvia, than in Armenia. This, can be explained, in part, by the fact that at an initial stage, the evolution of a country into a knowledge-based economy inevitably requires increased government participation to impact on the normalisation process by: coordinating and maintaining political, economic and legal mechanisms; creating conducive legislation and a normative-legal base; attracting investment and promoting competition; and encouraging the participation of the private sector by actively supporting national ICT producers. All these processes are taking place in all the states of Eastern Europe and CIS, but they are at different stages and levels, thus determining the level of readiness of each country for the knowledge-based economy.

- while many of the Eastern Europe and CIS countries have tremendous intellectual capital in ICTs, this can be affected negatively or positively by two conflicting processes. In some countries there is a significant erosion of intellectual capital due to unemployment, emigration, and the unstable situation in a number of ICT-oriented enterprises and universities. Conversely, however, there is a stabilisation and even a partial growth in this field due to the opening-up of new directions in ICTs and the development towards knowledge-based economies. Table 12, together with the country assessment reports, illustrates this fact. In countries of the high and even the middle groups, the latter, positive process is prevailing, while in all countries of the third group and in some countries of the middle group, the former, negative process is still dominant.

2.5. Technological trends

Having analysed global trends in ICT development and related issues, it is possible to move on to look at emerging trends in ICT technological change, as well as at the development of ICT services. Main technological trends are as follows:

- the last three to five years has seen a technological revolution as a new generation of computers has replaced the earlier one in all countries of Eastern Europe and CIS. However, the digital divide between the major subgroups of the countries not only persists, but is growing wider. Table 3 clearly illustrates this fact. In the countries of the highest subgroup, the number of computers per inhabitants is approximately 10 – 25% (Slovenia - 27.5%, Estonia - 17%, Latvia - 15%, Czech Republic - 12%, Hungary – 10% etc.). In the countries of the middle subgroup, this figure drops to approximately 5% (Russia - 5%, Bulgaria - 4.5%, Romania - 3.5% etc.). In the countries of the lowest subgroup, the number of computers per inhabitants is one percent or even less (Moldova - 1.5%, Armenia - 1% etc.). To put these figures in perspective, it is worth noting that in the USA the comparable statistic is more than 60%;

- expenditure on R&D differs from 0.2 – 0.5% in the countries of the lowest subgroup (Armenia – 0.1%, Kazakhstan – 0.2%) to approximately 1% in the countries of the middle subgroup (Russia – 1.1%) and to 1.5 – 2% in the highest country subgroup (Slovenia – 1.5 %). The number of people engaged in R&D, however, does not directly correspond to the budgetary resources allotted to R & D. For example, the Russian Federation has 3.4 % people in R&D, while in the Czech Republic this figure is 0.13%. This disproportion leads to substantial average salary differences in R & D. In the countries of the highest subgroup, the average salary for this sector is about US$ 1,000. In the countries belonging to the middle group, to which the Russian Federation also belongs, it is about US$ 100 – 200, while in the countries of the lowest group it is not more than US$30 – 40;

- software production is a leading segment of ICT production. There are three types of software producers in the countries of Eastern Europe and CIS. The first type comprises local companies who are mainly private and who produce different software programmes, primarily for local use. Some of these companies are comparatively large with up to 1,000 employers eg. IBS and IT in the Russian Federation. The second type of software producer comprises large international companies that have set up their branches and software houses in the region, and produce software for both international and local markets. Examples of this type include: Intel, Motorola, Sun Microsystems, Cisco, IBM, Telesoft Italy etc. The third type of software producer consists of small local companies or even individual programmers, working for "off- shore programming";

- hardware production is not growing so rapidly as software production. However, there are three main trends which can be mentioned. Firstly, the state sub-sector of ICT production is either decreasing or stagnating, but still produces most ICT products in Eastern Europe and CIS, especially, in the Russian Federation, Ukraine and Belarus. The average salary in the sub-sector is extremely low. Secondly, in recent years, a new sub-sector, consisting mainly of joint ventures between state enterprises and large foreign companies, has emerged. Examples are: Lucent Technologies and Siemens plants in St. Petersburg and Perm in the Russian Federation, and, in Chernigov and Kiev in the Ukraine; and PC and TV set production plants in many other countries in transition. The average salary in this sub-sector is comparatively high. Thirdly, another new sub-sector, consisting of very small and primarily private, small-sized enterprises and/or start-ups, has begun to produce its own equipment. Companies such as the Latvian company, SAF, and the Russian company, Natex, are characteristic of this sub-sector. Sole programmers, working mainly for offshore outfits, constitute another category of producers operating within this sub-sector;

- intensive growth of ICT services has caused an expansion in other areas of ICT production, eg. telecom accessories (amplifiers, connectors, cables etc). ICT services growth has also generated increased demand for specific software for the management, control and billing of telecommunications networks, adjusted to the needs of country operators and their customers;

- in many countries in transition, specialist companies have emerged locally, designing and integrating packages of software/hardware for large ICT projects, state-projects included. Some of these companies are comparatively large and powerful, and perform a buffering and bridging role between old state enterprises, or the old economy, and the "new" transition economy.

2.6. Trends in services

Major trends in ICT services and related issues in knowledge-based development in the countries of Eastern Europe and CIS are the following.

- ICT services and related sectors of the countries of Eastern Europe and CIS are growing more dynamically than ICT production itself. There are different reasons for this. Firstly, this is a global trend, not one which is taking place in countries in transition alone. Secondly, the ICT service sector has become a real driving force of economic development in countries in transition, a fact which is extremely important. While the production sector is stagnating, or only developing very slowly, in some of the transition countries, sub-sectors of ICT services, such as mobile communications and the Internet, are increasing at an annual rate of 20 – 50%. Thirdly, many large international ICT companies, which are presently experiencing a slump, now regard Eastern Europe and CIS as less risky and more profitable than before, and are choosing to become increasingly active in these countries;

- mobile communications are a special issue. The popularity of these is growing at a rapid rate even in some countries of the lowest group (see table 4). Countries such as Azerbaijan, Romania and Yugoslavia have reached an excellent mobile penetration rate (Azerbaijan – 8%, Romania – 17%, Yugoslavia – 19%) in comparison with other ICT services. In many countries of the region, mobile communications now constitute more than 50 to 60% of the telephone infrastructure as a whole, while even in the United States, this parameter is only around 40%. Rapid growth of mobile penetration is closely connected with new types of mobile services, such as prepaid, GPRS, info mobile etc. These respond to the needs and the poorer income levels typical of customers in the countries in transition;

- penetration of main telephone lines is taking place comparatively slowly. During the past five years, some countries, such as Armenia, Kazakhstan, the Kyrgyz Republic, Tajikistan, and Turkmenistan, have even lost some penetration (see table 3). In other countries, this parameter is growing, but not quickly as it is constrained by the poverty of the population. Another important issue is the comparatively low level of digital networks that allow the delivery to customers and final users of new enhanced telephone services. In most cases, these types of services are provided not by a country's own Telecom, but by other competitors, operators or carriers;

- some analysts and information sources show that this year has not been so good as regards Internet penetration in Europe. Prior to this year, international Internet capacity was, on average, doubling annually (with last year showing a three-fold increase). More, than 80% of all Internet international capacity is concentrated in Europe. However, during this year, there has been a 40% drop in this parameter. This is due to the fact that four very large international ISPs: Energis, Carrierl, KPNQwest and Teleglobe, were closed down (see www.telegeography.com) this year. In spite of this fact, the Internet in Eastern Europe and CIS is developing rapidly. Tables 3, 5, 6, 8 –11 illustrate this. Internet penetration is now is approximately 10 –30% in the countries of the highest subgroup (Slovenia – 30%, Hungary 15%, Czech Republic – 13%, Poland – 10% etc.). In the countries of the middle subgroup, Internet penetration is about 3 –7% (Bulgaria – 7%, Yugoslavia – 5.6%, Romania – 4.5%, Russia – 3%). The Russian Federation is viewed as a special case because Internet penetration is limited not only by economic level, but also by the size of the country. Internet penetration in the lowest group is about 1%, or even less. For the United States this parameter is about 50%;

- there is a clear downward trend in prices for mobile and Internet services. During the last two to three years, cost per minute for mobile telecommunications services fell by at least two times. The same has happened with dial-up charges per hour (see table 9). This is the only way in which operators, carriers and ISPs can gain a larger client base, especially within the poorer population segment;

- a revolution is taking place in broadband access to the Internet. Some three or four years ago, it was only possible for large or middle companies to access the Internet, using 64 –512 Kbps channels with committed information rate (CIR) and quality of service (Qos). Now, this service is available to small companies, SOHO and even, in some countries of the region, to households. This has taken place as a result of the rapid and wide spread of new types of broadband access, such as ADSL and wireless in some countries of the Eastern Europe and CIS, resulting in a drop in prices for final customers. For example, in the Russian Federation, during the last two to three years, prices for fixed broadband access to the Internet has fallen by three or four times;

- while in certain countries of Eastern Europe and CIS, the mobile communication services market is monopolised by two or three large operators, the Internet services market is more liberalised. In small countries in transition, there are tens of ISPs, and in large countries, there are even hundreds of them in strong competition with each other. This competition also creates a downward pressure on charges to customers as well as stabilising and even improving the quality of services;

- for transition countries, the ICT services sector is the most globalised part of their economy. Cross-border investments are increasing in this sector. Large international operators (Dutch Telecom, OTI, Tele 2, Equant, Cable&Wireless, Sonera, Golden Telecom etc.) work directly or through branches or partners in many countries of Eastern Europe and CIS, and, in many cases, are winning the competition with national operators or providers or carriers. For example, in the Russian Federation, the size of the alternative fixed communication market this year is about UN$ 700 million, one-fifth of the whole fixed telecommunications market of the Russian Federation;

- Internet use and e-services are growing at an unprecedented rate in the transition countries. Even in the USA, where Internet use and technology developed jointly, Internet penetration speed was still more than five times the speed of telephone penetration, and more than two times that of television penetration. Consequently, although Internet growth is taking place, the rate of growth of e-services is very unequal in the transition countries both as regards different subgroups, and among different population groups and economic sectors within individual countries. Ironically, the higher the growth rate, the wider the digital divide;

- information is one of the key factors that motivates people to use the Internet. Subsequently, absence of relevant content may act as a barrier to Internet access. The task of providing demand-driven information to all users is now one of the main driving forces for Internet penetration in the region. Determining the aggregated needs of individuals and defining where the existing Internet content fails or succeeds to meet these needs is both a global task and a marketing tool for many regional ISPs. The number of registered hosts per 10,000 inhabitants varies in different subgroups of the countries in transition. In the countries of the highest subgroup, this parameter is about 100 or 200 hosts per 10,000 inhabitants (210 – Czech Republic; 168 – Hungary; 126 – Poland). In the countries of the middle subgroup, this parameter is about 20 – 40 (33 – Bulgaria, 24 – Russia, Romania 21). In the countries of the lowest subgroup, this parameter is between one and five, and, in some cases, even lower (Belarus – 3,2, Azerbaijan – 1.3 and etc.). In the United States, this parameter is 3,714. Consequently, while the difference in the actual number of Internet users per inhabitants is only around two to four times more in the United States than in the most ICT developed countries of Eastern Europe and CIS, the difference in the total number of Internet hosts is far more dramatic, with up to 150 times more hosts in the United States. This illustrates to what extent the countries of Eastern Europe and CIS are behind in the fields of e-media, e-business and e-commerce applications. The infrastructure necessary to address this issue is only now being put into place.

2.7. Success stories

UNECE country assessment reports together with other sources, such as ITU and the World Bank, provide interesting stories that illustrate the progress being made by countries of Eastern Europe and CIS in their move towards a knowledge-based economy. This progress is often the result of joint efforts between governments, international organisations, and local and foreign companies.

2.7.1. Networked access

Latvian Export & Import Directory Project

The online information portal, "Latvian Export & Import Directory", was founded in December 2001. The project is aimed at Latvian companies interested in establishing export/import relations and attracting foreign investments. The portal provides information free of charge to Latvian and foreign companies. It offers Latvian companies the information necessary for their business development and amends information about their services and products in the Latvian exporters and importers database, formatting it for the target audience – foreign companies. The portal provides foreign companies with information about the producers, exporters and importers of different products and services in Latvia, and with information about business opportunities in the country. Business information is currently provided by the Latvian Development Agency

(LDA); the European Information Centre; the Ministry of Foreign Affairs; Embassies of the Latvian Republic; and associations of business enterprises. The development of the portal is still in progress and will be carried out in several steps. In the near future, it hopes to offer other marketing solutions to companies such as virtual product catalogues, homepages, etc.

2.7.2. Networked learning

Hungary's SuliNet

Launched in 1996 by the Ministry of Culture and Education, the SuliNet portal (now Irisz-SuliNet) (www.sulinet.hu, www.itu.org), was designed as a central repository of public education materials for teachers, students and parents throughout Hungary. With an initial budget of US\$ 11 million (1997) and sanctioned by an amendment to the Public Education Act, Irisz-SuliNet has proved to be a sustainable and productive response to the ever-changing decentralised public education system in Hungary. On the technical development side, the portal helps teachers to develop their IT skills, including infrastructure basics, network planning, organising and implementation. On the social side, it provides users with an e-mail account, list of services, a newsletter and a variety of other cultural and social development tools. With more than 2,000 active institutions in its network, Irisz-SuliNet has connected all the Hungarian secondary schools and more than 20% of primary schools to the Internet. In addition, Irisz-SuliNet is closely connected and supported by the agreement of the EU Heads of State and Government with the eEurope 2002 Action Plan, which was set out in June 2000 as a roadmap to achieving the eEurope initiative's targets.

Yugoslav Academic Network

The heart of the Yugoslav Academic Network is the Belgrade Academic Network, which belongs to Belgrade University. The Academic Network has over 100,000 users and more than 10,000 e-mail accounts. Computer labs are generally only open for computer studies during the day and closed after school, or may be open for teachers for class preparation but closed for students. All Universities in the country are connected to the Belgrade University Computing Centre at the speed of 2 Mbps. They are also connected to the Internet via a local commercial provider at the speed of only 512 Kbps. The existing 100 Mbps city optical backbone (Belgrade) connects two central nodes of the Belgrade Academic Network, as well as 63 Academic institutions. Telecom Serbia allocated 60 landlines to the University lecturers, for accessing the Network from their homes. More information about the Academic Network can be found on the net (http://servlet.rcub.bg.ac.yu/).

The "Chitalishta" project in Bulgaria

The "Chitalishta" project, sponsored by United Nations Development Program, USAID and the Dutch Government, is developed and supervised by the Ministry of Culture of Bulgaria and aims to encourage the cultural and educational role of traditional community centres. Under an agreement between UNDP and Cisco Systems, ICT training and special qualification programs will be implemented in a total of 25 Bulgarian "Chitalishta" and in three regional centres nation-wide.

2.7.3. Networked society

Tele cottages in Hungary

In a country that has over 3,000 small villages and where 7.8 per cent of the population lives in settlements with less than 1,000 people, Hungarian tele cottages are a key source of access to the global information society. Inspired by the tele centre schemes in Denmark and Sweden, a group of Hungarian librarians set out in 1990 to provide marginalised groups with computer services and Internet access. The first tele cottage, or "TeleHaz," in Hungary was established in a small mountain community called Czakbereny in 1994. For further information visit: http://www.itu.int/ITU-D/ict/cs/hungary/material/hungary.pdf.

Hungary's MEDINFO

In 1999, Hungary's Ministry of Health sought to leverage the power of new technologies to provide health care information for all Hungarians. By creating MEDINFO (www.medinfo.hu, www.itu.org), an online reference source and database that is the central node for health care-related communication in the country, the Ministry has made it easier for health care professionals to coordinate activities, while providing the public with a user-friendly interface to access the information they need to maintain mental and physical health. The site offers many different services (online catalogues, topic monitoring, literature searches etc.); public information for health care workers; a national library of heath science; a health line etc. The site also serves as a repository for information to resources outside of Hungary such as the World Health Organisation.

2.7.4. Networked economy

Project "Support to Industrial Cluster Restructuring" in Latvia

Project "Support to Industrial Cluster Restructuring", funded by the EU Phare, and lasting from October 2000 till October 2001, was the first cluster development initiative in Latvia. The aim of the project was to raise competitiveness within Latvian industry by popularising the concept of clusters and providing consultative support to potential individual clusters. Numerous areas of business activity were analysed in the initial phase of the project implementation, identifying fields where Latvia had opportunities to develop enterprise clusters. Four potential enterprise clusters were chosen for further consultative support. Identifying these by their basic activity, they could be grouped as follows: information systems clusters; forest clusters; composite materials clusters; engineering clusters. Three of these are characterised as hi-tech or knowledge based. One of the objectives of the project was to look for possible cooperation with enterprises of other sectors and research institutions, which in the future could potentially become the base for the creation of new innovative products.

Programming School in St. Petersburg, Russia and Booming Offshore Programming

St. Petersburg is one of the main centres of ICT training in Russia. According to expert estimates, Russian universities produce nearly 5,500 ICT specialists annually. Up to 1,500 of these are graduates of St. Petersburg universities (89% from five main schools). High levels of University education are strengthened by the strong emphasis on technical and fundamental sciences that was laid down in the Soviet period. Today, the main Universities have modern educational facilities. Moreover, increasing cooperation between leading universities and ICT companies has led to the growth of up-to-date educational programs. The quality of this education is renowned worldwide. The St. Petersburg State University team won first place at the International Collegiate Programming Contest in 2000 and 2001.

The concentration of a highly qualified workforce, international contacts, and a modern telecommunications infrastructure, have all helped St. Petersburg to become the main centre for offshore programming in the Russian Federation. In another big Russian IT centre, Moscow, programming business is more concentrated on public orders. Offshore programming in Russia is growing by 20% annually, amounting to about US$ 300 million in 2002. According to McKinsey Global Institute Research, the labour productivity in the Russian IT sector is 38% the USA level, but this is more than twice the Russian average. Certain cases, studied by McKinsey in Russia, have shown the labour productivity to be almost three times greater than the USA level (293%).

Support of local entrepreneurs in Bulgaria

The US Agency for International Development supports local entrepreneurs by providing technical assistance, limited hardware, software and Internet connectivity, as well as training and on-going technical and business management support to three selected PC telecenters in small, under-served Bulgarian towns. The PC telecentres train local users and grant free or subsidised access to specifically targeted social groups, women, minorities and the unemployed, in effort to address specific community needs and to create a customer base for telecenters.

2.7.5. Networked policy

State Automatic Election System Shailoo in the Kyrgyz Republic

A significant contribution to network software applications in the Kyrgyz Republic was made via the creation of Shailoo, a State automatic election system. Shailoo is a network-based software application designed to collect and process information at different stages for the republic's Election Committee. Client/server technology is used to collect data from 54 rayons (counties) over the whole country into the Oracle database of the Central Election Committee in Bishkek. A sophisticated management system protects the data from being manipulated at the electronic stage of information processing. During the Election campaigns, the results of the votes are shown in real time on the Internet. The system has been in operation since November 1998.

Computerised Taxation System in the Kyrgyz Republic

A computerised taxation system for State tax inspection was created by the Ministry of Finance with the participation of the Barents Group. This is a network-based software application, which uses communication channels of the Shailoo corporate network. A common information environment exists in the form of an Oracle database which is shared by the Tax Inspection Committee and the Social Fund. Branches in rural areas send data to Bishkek through Shailoo communication facilities.

Public Management Information System (PMIS) in the Kyrgyz Republic

This Internet-based State corporate network connects, through dedicated lines, 49 major Ministries/agencies in Bishkek and six regional State bodies in Osh. Existing telephone lines are used as communication channels, and network operations function up to transport layer. This allows e-mail and Internet access for all PMIS users.

The GIPI program in the Russian Federation

The GIPI (Global Internet Project Initiative) program in Russia (www-gipiproject.org) began work in earnest in mid-2001 when key senior staff were appointed. The main objective of GIPI Russia is to promote the adoption of a legal and regulatory framework that will support the development of open, decentralised, market-driven and transparent information and communications networks in Russia. GIPI Russia works to educate deputies of the State Duma and other policymakers on a non-partisan basis. GIPI also provides technical assistance on policy issues.

GIPI allowed for the summarising of legislative developments that occurred in the field of Internet regulation in 2001, as well as for the formulation of a 2002 work plan for the major Russian e-program, "Electronic Russia." GIPI Russia prepared and distributed a report, summarising the main legal developments concerning ICTs in 2001, "Bulletin of Legislative Activities in the Sphere of ICTs, 2001". It has been decided that GIPI should continue to collaborate with "Electronic Russia" in 2002 by providing a group of experts with an analysis of the proposed Internet and ICT legislation, preparing monthly updates on legislative developments in Russia and abroad, and organising further expert gatherings on "Electronic Russia" with the participation of outside experts on Internet regulation.

3. Capacity and e-readiness of the region for the Knowledge-based Economy

It must be recognised that the capacity and e-readiness of the transition and emerging market economies for the knowledge-based economy is not a static concept and evolves in response to many factors. These factors include: the level of economic development achieved; historic and cultural background; availability of different types of technology; and, particularly, the way in which a country's inhabitants interface with technology. For example, the average Internet user in Eastern Europe or CIS will, on the one hand, be considered more prepared for Internet use than people in other parts of the world, due to the high level of knowledge in these countries, but, on the other hand, as being relatively unprepared due to the comparatively low PC penetration levels. The same applies to readiness to work in ICT production. The potential is very high, but the diffusion of knowledge concerning the newest technologies throughout the population is, in some instances, very low. In both examples, special training and providing interfaces that allow such users to interact with new technologies is decisive in preparing the population for a knowledge-based economy.

The digital readiness of a country implies not only the availability of reliable and affordable ICT infrastructure, but also the following:

- availability of human capital that is capable of using, innovating and adapting new technologies;

- capacity of different sectors of the economy and society to accept and absorb ICT;

- creative knowledge as an independent force in the society;

- an organisational infrastructure to legislate and implement an appropriate legal framework for the use of ICTs in these sectors.

3.1. Availability of human capacity

Literacy is a basic instrument for the use of the Internet in its present form. Most countries in transition and emerging market economies have comparatively high adult literacy rates, and this factor creates additional advantages for them.

However, becoming a regular user of the Internet takes more than literacy. To overcome the difficulties associated with learning a new technology and to become familiarised with computer-based applications requires a digital or computer literacy. There are no direct ways of estimating the digital literacy rate of a country. However, proxy measurements, such as total PC penetration or PC penetration in schools give an indication of the level of digital literacy of the adult population or of the youth. The PC penetration per 100 inhabitants ranges for the measured countries of the region from 0.76 in Albania to 27.57 in Slovenia (see table 3). For the lowest group of the transition countries, typical PC penetration is lower than 1% (Albania – 0.76%, Armenia – 0.79%), for countries in the middle group it ranges from 3–6% (Romania – 3.57%, Russia – 4.97%, Bulgaria – 4.43%), and for the highest group it is more, than 10% (Latvia – 15.31%, Slovak Republic – 14.81%, Slovenia – 27.57%). For the United States, this parameter is 62.25%. PC penetration in schools might be measured in PCs per school (see table 10). It ranges from 0.225 in Armenia to 13.3 in Latvia. In the context of the countries in transition, where PCs are unaffordable for most households, the number of computers per school may be the more useful measurement.

To adopt, innovate and fully optimise ICTs, as an instrument of social and economic growth, digital and computer literacy is not enough. A country also needs a technical and scientific base. Table 16 is based on data from Harvard University and summarises the availability of technicians and scientists per million of population in different regions.

Table 16. Availability of human capital

No	World region	Technicians / million people	Scientists/ million people
1.	OECD	1326.1	2649.1
2.	Middle East	177.8	521
3.	East Asia	235.8	1026
4.	Latin America and Caribbean	205.4	656.6
5.	Eastern Europe and CIS	577.2	1841.3
6.	Sub-Saharan Africa	76.1	324.3
7.	South Asia	59.5	161

From these results, it is clear that the countries in transition and emerging market economies have extraordinary potential in science and technology, ranking second after the OECD countries. In spite of the importance of the human capita availability parameter, it shows only the intention for readiness towards a knowledge-based economy. For example, some countries from even the lowest subgroup have a high availability of such human capital (Armenia – 1,307, Azerbaijan – 2,735, Uzbekistan – 2,235). Some of the countries from the highest subgroup have even stronger figures (Estonia – 2,164, Hungary – 1,249) although some, such as Latvia, despite being in this group, have a low availability of human capital (Latvia - 361).

However, as seen in table 14, the average salary level in this sector is vital for sustaining and effectively utilising this human potential. Because of differentials in these areas, some countries are at risk of losing their human capital potential. In the countries of the lowest subgroup, the average salary in this sector is approximately US$ 30, in those belonging to the middle subgroup it is approximately US$ 100. In Latvia, however, the average salary in this sector is about US$ 1,000 explaining why Latvia, with a comparatively low availability of human capital, has been successful in moving towards a knowledge-based economy.

3.2. Capacity of different sectors

3.2.1. Networked Access

There are many factors that prevent potential users from benefiting from the Internet. The first of these is connectivity (the availability and reliability of the infrastructure to access the Internet). Second is the cost involved in accessing the Internet on a regular basis (affordability). Third is the substance and language of the content, i.e., relevance, comprehensibility, etc.

Both direct and indirect measures are used to assess the size of the Internet. The most common direct measures of Internet access are the following:

- number of hosts;
- number of users;
- number of subscribers.

A brief description of these metrics for the countries of the UNECE region and a summary of the merits and limitations of each of the metrics is presented in Tables 3, 6, 7, 9 in comparison with the United States.

Some of the most common indirect measures of Internet diffusion are:

- Tele-density;
- PC-density;
- Mobile-density;
- Cable TV density.

These measurements are presented in tables 3, 6, 7, 9 for PCs and in tables 4, 5, 8 for the other parameters.

Tele-density, given as the number of telephones per 100 people, is a necessary pre-requisite for dial-up access to the Internet, not including wireless access. Similarly PC-density, measured as the number of PCs per 100 people, is also a pre-requisite for all types of Internet access. Mobile density, on the other hand, measured as the number of cellular phones per 100 people, provides the upper limit for Internet access through wireless mobile technologies, such as WAP or GPRS. Cable TV density provides the upper bound of cable access to the Internet. Data from Harvard University shows that 15 countries have the highest number of users and the fastest rate of penetration (per thousand population). These fifteen countries (encompassing approximately 25% of the world population) accounted for 78% of the world's Internet users at the end of 1999.

In considering both direct and indirect country measurements, the United States is chosen for the purpose of comparison. When analysing the rate of Internet penetration growth for the most Internet-developed countries, it is obvious that the highest rate of growth will have taken place when penetration was between 10 and 50%. Tables in chapter 2 show that many countries of Eastern Europe and CIS have either already passed this point or are currently passing it. As such, a high growth of Internet penetration in the countries in transition in the near future can be predicted. Approximate calculations show that countries of Eastern Europe and CIS constitute approximately 5 –7% of world Internet users and 7 – 10% of global international traffic.

An additional factor may also promote future growth in some transition countries. Given that most Internet users also have PCs, the ideal situation is when all PCs are connected to the Internet. In the United States, PC penetration is about 60% while Internet penetration is about 50%, creating a difference of approximately 45%. Meanwhile, in Latvia, PC penetration is approximately 15% while Internet penetration is approximately 7%. In Russia, PC penetration is about 5% and Internet penetration is about 3%. In the cases of these transition economies, therefore, the difference is almost 100%.

In the United States, the ICT sector is very large in absolute terms and is practically balanced in that PC and Internet penetration are almost equal. This is very good for the US economy, but not for the future growth of the sector. More than 50% of the population is already using the Internet, and in every society there will always be some people who will never become users of this technology. On the contrary, in many countries in transition, PC and Internet penetration rates are comparatively low, leading to an expectation of rapid growth in the near future bringing the economies of Eastern Europe and CIS closer to a knowledge-based economy. However, as noted, the parameters are very unbalanced, with PC penetration often several times higher than Internet penetration. This can be seen as an extra stimulant to growth since PC users are primed to become Internet users, the only problems being those of income and accessibility (affordability and availability).

There are many other indicators for networked access, such as: quality of service; opportunities for collective access; cost of access; and penetration of local content both within the country itself and outside. All these indicators are analysed in the UNECE country assessment reports, while some are presented in tables in chapter 2. Results show that, over the last two to three years, there has been a real revolution in price and quality of services not only in the highest subgroup of countries in transition, but also in the middle subgroup, demonstrating a further possibility of rapid growth in the near future.

3.2.2. Networked Learning

The term "networked learning", by its more narrow definition, covers all types of learning and educational processes being supported with ICTs. In its wider definition, however, this term means the use of ICTs, PCs, and the Internet in all segments of education, training, learning, and certification. There is a general tendency towards the transformation of traditional learning into a continuous life-long process, making networked learning an important indicator of a country's readiness for the knowledge-based economy. In comparison with other readiness indicators "networked learning" helps in understanding and predicting the five to ten years ahead.

A common proxy approach to assessing the level of networked learning in a country is to analyse PC penetration in education as a whole and in its different sectors. Table 7 provides data on the total number of PCs in education for the eleven countries of Eastern Europe and CIS. This information was taken from the World Bank database. Table 10 presents the same indicators, derived from the country assessment reports.

From the information provided in table 7, it becomes obvious that almost all measured countries are at the same proxy level, with the exception, perhaps, of Poland which shows a slightly higher numbers of PCs installed for education. In comparison with the United States, the scenario from this table is bleak for the transition countries with the difference between them and the USA amounting to 100-200 times, even allowing for recalculations by inhabitants or by schools. When taken together with the registered number of domains, the gap between the USA and the countries in transition appears even wider.

However, as illustrated in table 10, the situation in some of the nine countries covered by the country assessment reports, such as Bulgaria, Latvia and Belarus, is improving slightly. Success stories from chapter 2, such as SuliNet in Hungary, Yugoslav Academic Network, and the "Chitalishta" project in Bulgaria, demonstrate that there are many successful attempts to improve the situation in networked learning. The problem is the low average level.

Digital divides exist between countries in transition, emerging market economies and the most developed countries, between some of countries in transition and emerging market economies of the UNECE region, between different stages of education (high, secondary and primary), and between urban and rural regions. In order to overcome these digital divides, which are sometimes even growing wider, networked learning needs to become one of the major development priorities of the transition and emerging market economies.

In spite of the above, viewing "networked learning" purely from its more narrow definition, would suggest that the necessary ground is being covered. ICTs exist as an independent sector of industry in all countries of Eastern Europe and CIS, and the problem of preparing human recourses for this sector is being solved by the sector itself.

3.2.3. Networked Society

Other sections of this chapter analyse existing disparities in Internet usage at a national level. However, it is important to pay equal attention to the differences within countries. As mentioned below, there are significant sub-national disparities. There are two approaches to alleviating these disparities. Firstly, these differences signify a divide requiring an active State intervention to resolve them. Secondly, any observable disparity does not signify a rigid division but the difference in the rates at which different population segments assimilate new technologies. These disparities are transient and with time will dissolve by themselves without external intervention. The following section investigates the validity of these approaches by examining the available data. Following are some of the critical sub-national dimensions of digital disparities:

- Income;
- Gender;
- Age;
- education level;
- geographic location of the user.

Income level seems to be a critical factor in determining Internet access. Data from the World Bank and Harvard University show that in some income groups, the penetration rate of Internet usage is approximately seven times larger than in the lowest income groups.

Studies on the United States show that the correlation between income level and Internet access is becoming weaker, evidencing that the Internet penetration rate for all income levels is rising. At the lowest income level (under $15,000), the penetration level increased from 7.1% in 1998 (December) to 12.7% in 2000 (August), almost a 79% increase. During this same period, Internet penetration at the highest income level (above $75,000) increased from 60.3% in to 77.7% (a 17% increase). Between December 1998 and August 2000, growth in Internet penetration rates in the United States were observed to be higher at lower income levels as compared to growth in penetration at the higher income levels. However, the disparity in penetration between the highest income level and the lowest income level increased from 53.2% to 65%, during this period.

In Eastern Europe and CIS, the situation is different. The share of the lowest income population (for example, in Russia with a monthly average income less than US$ 70) using the Internet, is comparatively larger (43 % of all Internet users in Russia). This is quite understandable, because a comparative part of the low income population is simply larger in the countries in transition, and the share of the higher income population is not large enough to enable the growth of the Internet, which the economy needs.

Gender division among Internet users is also very important, and a gender disparity in access to the Internet is present in many countries throughout the world. However, it seems that the Internet has been an equalizing force and, in those countries, where no gender discrimination in education is being practiced, the proportion of Internet users in respective gender total populations is close to parity. According to some recent studies, some UNECE member countries are comparatively more advanced than others in terms of gender equity (see table 17).

Table 17. Internet users in select countries by gender, 2002.
(percentage of total male and female population)

Country	Men	Women
Belgium	53	47
Bulgaria	9	9
Canada	61	59
Czech Republic	32	28
Denmark	68	58
Estonia	36	41
France	43	31
Finland	59	60
Germany	46	36
United Kingdom	45	33
Hungary	15	7
Ireland	46	46
Israel	50	35
Italy	50	26
Latvia	17	17
Lithuania	19	18
Netherlands	68	54
Norway	66	50
Poland	19	15
Romania	16	8
Slovakia	27	21
Spain	34	23
Turkey	18	23
Yugoslavia (Serbia only)	18	14
Ukraine	6	2
United States of America	61	62

Source: eCommerce Internet Surveys & Online Market Research (http://www.tnsofres.com/ger2002/keycountry)

The United States, where more women than men are found to use the Internet, is an exception in this respect. The situation can be largely attributed to a pro-active Government policy and to the women's movement. The achievements of the United States were predetermined by several factors:

- high literacy and digital literacy rates of women;
- equal status of women in the social structure, and, correspondingly, the availability of relevant content;

- the affordability of Internet access, and women's ability to control their disposable income.

Therefore, gender disparities in access to the Internet are not inevitable, as the experience of the United States evidences.

Age disparities as regards access to the Internet are very noticeable in countries in transition. In several countries of Eastern Europe and CIS, older people are much less likely to access the Internet than younger people. For example, Russians in the age category of 18-24 years old are five times more likely to access the Internet than those above the age of 55 years old. The average age of Internet users in the United States is 36 years old, while, in countries of Eastern Europe and CIS, it is about 30 years year old. This is not surprising, however, as, in many countries with economies in transition, the younger generations are becoming digitally literate at school, and, also face fewer problems in changing career patterns than their elder counterparts did.

The importance of **the level of education** in accessing the Internet is easy to analyse given that historically the Internet appeared as a medium for communication between academic institutions (and military institutions) in the United States. Before the1990s, it was overwhelmingly populated by techno-literate college graduates. With the introduction of the worldwide web and commercialisation of access to the Internet in the early 1990s, the user base has somewhat broadened. However, the preponderance of highly educated users persists globally.

In Yugoslavia, one of the largest Internet networks is the network of Academy of Science (see chapter 2 success stories), the same is in Romania. There are very large Academic networks in Moscow, Russia (www.radionet.iitp.ru for example). In the Russian Federation, more than one third of Internet users are qualified specialists with higher education (36.7 %). Students, pupils and top managers are also active users of the Internet (27.5 %). The situation is the same in almost all the countries in transition.

Education level contributes to disparities in Internet use in three ways. The level of education often correlates with digital literacy and social capital. It also correlates with income and the ability to afford Internet access.

Harvard University data provides some evidence concerning the complex interaction between these three factors. Consider households headed by those with no high school education. Internet access rate depends on the level of income and ranges from less than 5% when annual income is less than US$15, 000, to 51% when income is above US$ 75,000 per annum (while the national average rate of Internet penetration for householders with this level of education is 11.7 per cent). However, the household income alone does not explain the impact of education. If we consider households with income levels of more than US$ 75,000 per annum (with average penetration rate of 77%), penetration rates of households headed by individuals with no high school education is 51%, while for those headed by individuals with college degrees it is 82%. This 31% spread clearly shows the role of digital literacy and social capital.

Internet/ICT penetration is also influenced by the **Geographic location** of the country itself and by the location within the country. Urban centres generally have better infrastructure and higher levels of income compared to rural areas. In many countries of Eastern Europe and CIS, the income level of different regions within the country, the level of ICT infrastructure development, the level of ICT service development and, consequently, the level of development of the knowledge-based economy are in direct correlation.

Countries of the lowest subgroup normally have Internet presence in only one or two cities, while countries of the high and middle subgroups have dialup Internet access nationwide. Broadband access, even in the middle and highly developed countries of Eastern Europe and CIS, is concentrated in major cities. A typical situation is as follows: the country's capital offers many opportunities for accessing the Internet, while in other cities, there are a few points of access. Furthermore, in all the countries, main Internet traffic is concentrated in cities. For example, in the Russian Federation, 36% of the Internet traffic is concentrated in Moscow and 9% in St. Petersburg; in the Slovak Republic, 60% of the Internet traffic is concentrated in large cities, 37% in small and middle cities, and only 3% in rural areas.

In many countries of Eastern Europe and CIS, broadband wireless access has become a main instrument for penetration of rural areas and, even, cities. There are two reasons for this. Firstly, due to poor infrastructure, wireless access, in many cases, may be the only solution. Secondly, possibilities for being serviced by alternative providers are not yet available in many countries (see the country assessment reports).

3.2.4. Networked economy

Many attempts have been made to provide an inter-country ranking of e-business readiness. McConnell International evaluated 42 countries for their e-business readiness by considering following indicators:

- connectivity (Are networks easy and affordable to access and to use?);
- e-leadership (Is e-readiness a national priority?);
- information security (Can the processing and the storage of networked information be trusted?);
- human capital (Are the right people available to support e-business and to build knowledge-based society?);
- e-business climate ("how easy is it to do e-business today").

On each of the criteria, countries were rated in terms of three grades: ready, improvements needed, or substantial improvements needed to become e-ready.

Among the European countries, Ukraine, the Russian Federation, Romania and Bulgaria show the least e-readiness, while Greece, Hungary and Italy showing the most e-readiness of the country sample studied.

The results of World Bank ranking were, however, different (see table 6). Eleven of the Eastern European and CIS countries were ranked from 1 to 7 (7 – the highest). The sample included countries from the highest and the middle subgroups. The results were between 4.5 (Estonia and Czech Republic) and 2.8 (Russia). Even Ukraine, far from being a leader in these areas among the transition and emerging market economies, had 4.1. What was also important, was that the United States was ranked only 5.

The results of the UNECE country assessment reports also show that in some areas of e-economy development, such as e-leadership and human capital, many countries of Eastern Europe and CIS are in a good position. Connectivity, in many countries, especially in the high and middle country subgroups, is quite good by international standards. The problems that need to be addressed are mainly in the areas of information security and e-business climate.

Many analysts and experts confirm the existence of a three-tier world system in e-economy. The greatest beneficiaries of e-economy are the countries of Western Europe, USA, Canada, Japan, Korea etc. There is a middle tier of countries capable of adopting new technologies and benefiting from them in the future. There is also a tier of low-income countries which is being bypassed by the ICT revolution.

This report recognizes that there is no quick fix for the structural changes needed in order to enable many low-income countries of Eastern Europe and CIS to improve their e-readiness, eg. digital literacy that includes both computer basics and techno-digital literacy. However, some of the UNECE transition and emerging market economies have already moved into the second tier.

3.2.5 Networked Policy

Potential benefits for ICTs and networking in the Government sector are enormous. They can have a significant impact on the shaping of institutions and communications networks of governments with the private sector and civil society. ICTs enable government bodies to work with greater transparency (from decision-making processes and implementing policies to awarding contracts) and accountability (by providing citizens with better access to information regarding their rights and benefits), and ensure collective participation (by providing a direct communication channel between the Government and citizens/residents of the nation).

Increasingly different state websites have gone online in many parts of the world. Table 18 summarizes for each country in transition, the total number of government websites that have gone online. The Harvard University data, although not very recent, gives some indication of the ongoing process of e-government development in the countries under consideration (2000).

The countries are ranked by the total number of sites, with the United States at the top of the list for comparison. The list includes: state bodies; political parties; judiciary; legislature; etc. Since the basic functions of the government remain similar across all the countries, the total number of government sites online is relevant rather than a population-adjusted measure.

As can be seen from table 18, some countries from the middle subgroup, such as the Russian Federation and Romania, are very successful in this sector. Data in brackets is has been taken from the UNECE country assessment reports (table 11). In some cases, indicators for the current year are even lower.

Table 6 also gives a ranking by this indicator of many countries of Eastern Europe and CIS. The primary results of this World Bank ranking are in concordance with the results presented in table 18.

Table 18. Government Websites

No	Country or Territory	Number of sites	File size (in MB)	Date of last change
1.	The United States	828	84	8/23/00
2.	Poland	112	22	8/27/00
3.	Russian Federation	104 (70)	17	8/29/00
4.	Estonia	103	16	8/24/00
5.	Czech Republic	83	15	6/24/00
6.	Slovenia	82	16	8/22/00
7.	Hungary	80	17	8/31/00
8.	Croatia	60	12	6/22/00
9.	Romania	60	14	8/31/00
10.	Latvia	54	11	7/15/00
11.	Slovak Republic	48	11	8/26/00
12.	Yugoslavia	48	12	8/29/00
13.	Lithuania	47	10	1/22/00
14.	Bulgaria	46 (150)	9	7/30/00
15.	Bosnia	35	10	8/26/00
16.	Ukraine	30	8	7/18/00
17.	Macedonia	27	8	7/28/00
18.	Kazakhstan	17	7	8/6/00
19.	Armenia	16 (10)	6	8/30/00
20.	Republic of Belarus	16 (55)	7	8/4/00
21.	Kyrgyz Republic	16	7	8/2/00
22.	Georgia	15 (10)	6	8/15/00
23.	Albania	14	7	8/2/00
24.	Uzbekistan	13	6	5/26/00
25.	Azerbaijan	12	6	8/28/00
26.	Moldova	12	6	8/31/00
27.	Turkmenistan	3	5	8/2/00
28.	Tajikistan	2	5	6/4/00

3.3. Investment in information infrastructure and R&D

As mentioned in chapter 2, a long-term trend of connectivity in any country is dependent on its current efforts to build a nationwide information infrastructure, and on ongoing Research and Development in ICTs.   Spending on ICTs in general, and on investments in information infrastructure and the R&D sector in particular, provide reasonable proxy measures for the long-term prospects of growth in connectivity. The results for Eastern Europe and CIS in this sphere are presented in tables 2 (World Bank) and 13 (from the country assessment reports). The measurements of the World Bank are for 2001, while the measurements of the country assessment reports are for 2002.

In terms of aggregate spending on ICTs, many middle-income countries in transition and emerging market economies are growing at rates higher than those of the established ICT players in the world. For example, ICT spending in Latvia has grown by more than 80% in 2001-2002. The comparable rate for Western Europe is 6.2% (World Bank).

The findings of some less-recent studies conducted by the World Bank for 59 countries are summarised in table 19. Significant disparities are apparent in investment (both in R&D and in information infrastructure) between the high-income countries and their low-income counterparts. From this study, it is obvious that in 2000 the transition and emerging market economies all together were only in fifth place among the other world regions for investment to the information infrastructure, but were previously in second place for investment to R&D. The data in table13 shows that, in some countries of Eastern Europe and CIS, the present situation is slightly better.

In summary, the current, wide disparity in the investment pattern between the poorest and the richest countries inside and outside of Eastern Europe and CIS, implies that the existing disparity in access between the two ends of the spectrum is likely to grow, rather than diminish in the near future. This is a worldwide tendency. Similar data on the spending patterns of the middle-income countries of the world (to which many countries of the high and middle subgroup of the Eastern Europe and CIS belong) suggests that the gap between the middle of the spectrum and the high-income countries may decline as a result of sustained higher investments by the former.

Table 19. Investment to information infrastructure and R&D

Region	Investment in information infrastructure (per capita, USD)	R&D as % GDP
	2000	**1992-1997**
OECD	94.38	1.8
Middle East	24.57	0.4
East Asia	26.07	0.8
Latin America and Caribbean	38.78	0.5
Eastern Europe and CIS	18.96	0.9
Sub-Saharan Africa	7.92	0.2
South Asia	4.67	0.8

4. Indexing of the transition and emerging market economies

4.1. ICT indexing

There are many different approaches to indexing or ranking different countries. Composite indexes of different parameters of production and services, such as the Global ICT Index, have been developed in an attempt to measure and present by one number a country's relative position in the production or use of ICTs. Such figures are also indicative of the level of a country's development towards a new economy. Most approaches use a special ranking procedure, based on normalised indexes of different sectors.

The ITU employs one such approach. The conceptual and mathematical content for this can be read on www.itu.org under Global ICT Index. The idea behind the ITU approach is as follows: the index shows which economies are poised to take advantage of ICTs from the point of view of e-readiness. E-readiness is regarded as a measure of current ICT development and as a measure of an economy's ability to take advantage of ICTs in the future. E-readiness components include: infrastructure; usage; and market structure.

The infrastructure component consists of six variables (main lines, cellular subscribers, estimated Internet users, personal computers, international Internet bandwidth and broadband subscribers) and is included in indexing if three of these variables are present.

The usage component consists of six variables and is also included in indexing if three of these variables are present. These variables include: total national telephone traffic; international outgoing and incoming telephone traffic; telecom revenue; rates per minute with the United States; and the cost of a three minute local call).

The market structure component consists of ten variables and is included in indexing if five of these variables are present. These variables include: privatised incumbent; number of years incumbent has been private; independent regulator; number of years regulator has been independent; separate market structure for local services; long distance international calls; mobile service; leased lines; and ISPs .

All individual variables are normalised to one, so that they can be combined into a composite variable. The variables of each group are aggregated to form the three index components from which the global index is calculated.

4.2. Global Competitiveness indexing

There are also many different approaches to indexing and ranking the extent to which individual national economies and their structures, institutions, and policies are ready for economic growth in the medium term using new technological approaches. One approach is described in The World Economic Forum Highlights of the Global Competitiveness Report 2001 – 2002. The idea of this approach is to index national economies by the Growth Competitiveness Index (GCI). This is comprised of three sub-indexes: technology; public institutions; and macroeconomic environment. The technology index measures the capacity for innovation and diffusion of technology. The public institutions index primarily measures the role of politics and the bureaucracy in supporting market-based economic activity and the division of labour. The macroeconomic environment index measures variables related to capital accumulation and the efficiency of the division of labour.

Recognising that technology plays a different role at the various stages of economic development, the last report separates all countries into two groups based on their level of technology advancement. Using patenting as a measure of innovative capacity, the GCI identifies 21 innovation-driven economies, termed as the core economies. The other non-innovating economies are termed the non-core economies.

For the core economies, there is statistical evidence that innovation plays a dominant role in the countries with the medium growth rate, as they have already attained a minimum level of macroeconomic and public institutional progress. For these economies, the GCI thus places a weight of 1/2 on the technology index against weights of 1/4 each on public institutions and macroeconomic environment. The focus on technology is also weighted by the capacity to innovate.

Global Competitiveness Index

Core GCI = 1/2 technology index + 1/4 public institutions index + 1/4 macroeconomic environment index,
where

Core Technology Index = 1/2 innovation sub-index + 1/2 ICT sub-index.

4.3. Knowledge-Based Economy indexing

In this report, we suggest the indexing of a country's readiness for a knowledge-based economy by the Global Knowledge-Based Economy Index (GKEI). The calculation of GKEI takes into account all previous experience (see paragraphs 4.1 and 4.2 above) and some general ideas about knowledge-based economy indexing. The ideas are as follows:

- GKEI, like GCI, is comprised of three sub-indexes: technology (TGKEI); public institutions (PGKEI); and macroeconomic environment (MGKEI).
- the knowledge-based economy is regarded as the core economy and because of it the weights of sub-indexes are variable and flexible, depending on parameters included in the calculation of each sub-index, component and coefficient.
- Each sub-index, or component of a sub-index, or coefficient is normalised by the same sub-index, or component, or coefficient for the United States.
- The largest index is the best. .
- We are suggesting only a direct calculation of GKEI without any additional ranking.
- Below, we will try to evaluate GKEI by proxy measures available from this report.

The Global Knowledge-Based Economy Index is calculated using the formula below:

Global Knowledge-Based Economy Index

$$GKEI = A \ TGKEI \ + \ B \ PGKEI \ + \ C \ MGKEI, \tag{1}$$

where

A, B and C are weight coefficients and $A + B + C = 1$.

We'll assume $A=1/3$, $B=1/6$, $C=1/2$.

The Technology Global Knowledge-Based Economy Sub-index TGKEI is calculated by following formula:

$$TGKEI = 1/5 \ NAC \ + \ 1/5 \ NLC \ + \ 1/5 \ NSC \ + \ 1/5 \ NEC \ + \ 1/5 \ IC, \tag{2}$$

where

- NAC – networked access component.

- NLC – networked learning component,

- NSC – networked society component,

- NEC – networked economy component,

- IC – innovation component.

Networked access component NAC is calculated by following formula:

$$NAC = (1/4 \ NIUP + 1/4 \ NIDP + 1/4 \ NMP + 1/4 \ NMLP) \ [1/SQR(NDC)], \tag{3}$$

where

- NIUP – normalised Internet users penetration coefficient from Table 3,

- NIDP – normalised Internet domain penetration coefficient from Table 3,

- NMP – normalised mobile penetration coefficient from Table 5,

- NMLP – normalised main lines penetration coefficient from Table 4,

- NDC – normalised population density coefficient from Table 1.

The purpose of the NDC is as follows: If the same results for the technology are achieved by different countries with different population density, then the country with the lowest density is receiving some additional gain, because the ICT infrastructure is covering a larger territory, so it is more expensive and more comprehensive.

This is a general approach to calculating GKEI. For the United States, the GKEI is equal to one. There are many variants for formulating the approach to the calculation of every sub-index, component and coefficient. The final calculation must result from a comprehensive measurement. Here we suggest a proxy quick calculation, where every coefficient, component or sub-index is calculated with a very small number of parameters, for which indicators are available from, amongst other sources, the UNECE country assessment reports.

The networked learning component, NLC, as a proxy estimate is calculated as a normalised number of PCs in schools per school from table 10, or as a normalised number of PCs installed in education from table 7. For countries, not represented on this criterion in tables 7 and 10, NLC is approximated by a normalised Internet domain penetration coefficient, NIDP.

The networked society component, NSC, and the networked economy component, NEC, can be a proxy estimate of a normalised number of registered domains in social and business spheres. Here, as an extra proxy estimate, we use also a normalised Internet domain penetration coefficient, NIDP, for NEC and a normalised estimate of the PC penetration from table 3 for NSC.

The innovation component, IC, is calculated as a normalised number of scientists and engineers in R&D from table 2. For some countries, for which this parameter is not available, we are using other parameters from table 2, such as a normalised total ICT as percentage of total GDP or expenditure on R&D. For Georgia, we are using the same parameters from the country assessment report from table13. For five countries, Tajikistan, Turkmenistan, Uzbekistan, Albania and Bosnia, we failed to find any parameters on measuring the innovation component. So, these countries are measured by the middle rank of neighbouring countries and are presented in the tables below.

The Public institution knowledge-based economy sub-index, PGKEI, is calculated as a normalised number of online Government sites from table 18.

The Macroeconomic environment knowledge-based economy sub-index (MGKEI) is calculated as a normalised number of GDP per capita from table1.

Once again, all normalisation operations are done to the same parameter of the United States.

4.4. Results and interpreting of indexing

Table 20 presents the results of the calculation of the networked access component for countries in transition and emerging market economies, for which data is available.

Table 20. Calculation of the Networked Access Component, NAC

No	Country	NIUP	NIDP	NMP	NMLP	NDC	NAC
1.	Armenia	0.0284	0.00168	0.0149	0.168	4.065	0.0471
2.	Azerbaijan	0.00643	0.000455	0.179	0.168	2.903	0.211
3.	Georgia	0.00903	0.00102	0.121	0.239	2.516	0.119
4.	Kazakhstan	0.0123	0.00183	0.0815	0.170	0.194	0.192
5.	Kyrgyz Republic	0.0212	0.00246	0.0122	0.116	0.807	0.0682
6.	Tajikistan	0.00105	0.000132	0.000675	0.0546	1.387	0.0134
7.	Turkmenistan	0.00331	0.000902	0.00473	0.121	0.323	0.064
8.	Uzbekistan	0.0119	0.0000215	0.00563	0.0990	1.807	0.0314
9.	Albania	0.00504	0.000127	0.199	0.0747	4.452	0.213
10.	Belarus Republic	0.0825	0.000864	0.0304	0.420	1.581	0.197
11.	Bosnia	0.00938	0.00215	0.129	0.167	2.581	0.166
12.	Bulgaria	0.149	0.00894	0.430	0.541	2.355	0.184
13.	Croatia	0.112	0.0127	0.849	0.550	2.645	0.234
14.	Czech Republic	0.273	0.0565	1.483	0.563	4.194	0.290
15.	Estonia	0.602	0.0961	1.025	0.530	1.032	0.554
16.	Hungary	0.297	0.0452	1.121	0.563	3.452	0.273
17.	Latvia	0.145	0.0286	0.629	0.464	1.194	0.290
18.	Lithuania	0.136	0.00956	0.570	0.471	1.807	0.221
19.	Moldova	0.0274	0.00108	0.108	0.232	4.194	0.0442
20.	Poland	0.197	0.0342	0.586	0.384	4.00	0.150
21.	Romania	0.0835	0.00557	0.388	0.275	3.032	0.115
22.	Russian Federation	0.0587	0.00650	0.0853	0.366	0.290	0.240
23.	Slovak Republic	0.241	0.0361	0.895	0.433	3.549	0.213
24.	Slovenia	0.602	0.0399	1.710	0.604	3.194	0.4175
25.	TFYR Macedonia	0.0685	0.00342	0.246	0.397	2.548	0.112
26.	Ukraine	0.0239	0.00312	0.0995	0.319	2.677	0.0681
27.	Yugoslavia	0.113	0.00405	0.421	0.344	3.387	0.120
28.	**United States**	**1.00**	**1.00**	**1.00**	**1.00**	**1.00**	**1.00**

Table 21 shows the results of calculations for the Technology Global Knowledge-Based Economy Sub-index

Table 21. Calculation of the Technology Global Knowledge-Based Economy Sub-index, TGKEI

No	Country	NAC	NLC	NSC	NEC	IC	TGKEI
1.	Armenia	0.0471	0.00168	0.00168	0.00168	0.319	0.108
2.	Azerbaijan	0.211	0.000455	0.000455	0.000455	0.367	0.115
3.	Georgia	0.119	0.00102	0.00102	0.00102	0.180	0.0604
4.	Kazakhstan	0.192	0.00183	0.00183	0.00183	0.120	0.0635
5.	Kyrgyz Republic	0.0682	0.00246	0.00246	0.00246	0.140	0.0431
6.	Tajikistan	*0.0134*	*0.000132*	*0.000132*	*0.000132*	*0.12*	*0.0267*
7.	Turkmenistan	*0.064*	*0.000902*	*0.000902*	*0.000902*	*0.12*	*0.0373*
8.	Uzbekistan	*0.0314*	*0.0000215*	*0.0000215*	*0.0000215*	*0.12*	*0.0302*
9.	Albania	*0.213*	*0.000127*	*0.000127*	*0.000127*	*0.0944*	*0.0616*
10.	Belarus Republic	0.197	0.000864	0.000864	0.000864	0.559	0.152
11.	Bosnia	*0.166*	*0.00215*	*0.00215*	*0.00215*	*0.0944*	*0.0534*
12.	Bulgaria	0.184	0.00894	0.00894	0.00894	0.314	0.105
13.	Croatia	0.234	0.0127	0.0127	0.0127	0.364	0.127
14.	Czech Republic	0.290	0.0565	0.0565	0.0565	0.322	0.156
15.	Estonia	0.554	0.0961	0.0961	0.0961	0.527	0.274
16.	Hungary	0.273	0.0452	0.0452	0.0452	0.304	0.173
17.	Latvia	0.290	0.0286	0.0286	0.0286	0.088	0.154
18.	Lithuania	0.221	0.00956	0.00956	0.00956	0.496	0.249
19.	Moldova	0.0442	0.00108	0.00108	0.00108	0.0814	0.0258
20.	Poland	0.150	0.0342	0.0342	0.0342	0.356	0.116
21.	Romania	0.115	0.00557	0.00557	0.00557	0.278	0.100
22.	Russian Federation	0.240	0.00650	0.00650	0.00650	0.827	0.362
23.	Slovak Republic	0.213	0.0361	0.0361	0.0361	0.416	0.148
24.	Slovenia	0.4175	0.0399	0.0399	0.0399	0.527	0.213
25.	TFYR Macedonia	0.112	0.00342	0.00342	0.00342	0.0944	0.0433
26.	Ukraine	0.0681	0.00312	0.00312	0.00312	0.617	0.139
27.	Yugoslavia	0.120	0.00405	0.00405	0.00405	0.582	0.143
28.	**United States**	**1.00**	**1.00**	**1.00**	**1.00**	**1.00**	**1.00**

Table 22 shows the results of calculating the Global Knowledge-Based Economy Index. The country with the largest GKEI is the first, while the country with the lowest is last.

Table 22. Calculation of the Global Knowledge-Based Economy Index, GKEI

No	Country	TGKEI	PGKEI	MGKEI	GKEI
1.	United States	1.00	1.00	1.00	1.00
2.	Slovenia	0.213	0.0990	0.252	0.214
3.	Russia	0.362	0.125	0.0472	0.164
4.	Estonia	0.274	0.124	0.0954	0.160
5.	Czech Republic	0.156	0.100	0.136	0.137
6.	Hungary	0.173	0.0966	0.126	0.137
7.	Lithuania	0.249	0.0568	0.0840	0.135
8.	Poland	0.116	0.135	0.113	0.118
9.	Croatia	0.127	0.0725	0.118	0.117
10.	Slovak Republic	0.148	0.0580	0.0978	0.108
11.	Latvia	0.154	0.0652	0.0810	0.103
12.	Yugoslavia	0.143	0.0580	0.0295	0.0721
13.	Romania	0.100	0.0725	0.0452	0.0680
14.	Belarus	0.152	0.0193	0.0225	0.0652
15.	Bulgaria	0.105	0.0556	0.0407	0.0646
16.	Ukraine	0.139	0.0362	0.0168	0.0607
17.	Azerbaijan	0.115	0.0145	0.0142	0.0486
18.	Armenia	0.108	0.0193	0.0150	0.0467
19.	TFYR Macedonia	0.0433	0.0326	0.0471	0.0434
20.	Bosnia	*0.0534*	*0.0423*	*0.0325*	*0.0411*
21.	Kazakhstan	0.0635	0.0205	0.0269	0.0381
22.	Albania	*0.0616*	*0.0169*	*0.0260*	*0.0363*
23.	Georgia	0.0604	0.0181	0.0145	0.0304
24.	Uzbekistan	*0.0302*	*0.0157*	*0.0187*	*0.0220*
25.	Kyrgyz Republic	0.0431	0.0193	0.00704	0.0211
26.	Turkmenistan	*0.0373*	*0.00362*	*0.0161*	*0.0211*
27.	Tajikistan	*0.0267*	*0.00242*	*0.00492*	*0.0118*
28.	Moldova	0.0258	0.0144	0.00812	0.0151

Looking at the results in table 22, it is important to understand that they are proxy and are based on a comparatively small number of parameters, some of which date from 2000. This is a suggestion for one approach to calculate the Global Based-Economy Index and an illustration of how this can be used.

Again, the components of networked learning, networked society, networked economy and networked policy are an approximate proxy. It is absolutely clear that this is only the beginning of a long and comprehensive work. Also, the question of how to calculate the networked access component remains challenging. Herein, we suggest an approach, including in the use of country population density. The reason for this is obvious. For large countries with small population density, investment in developing a networked infrastructure (backbone) is larger. Due to this approach, the Russian Federation with a very small population density has received a very large additional gain and has moved from 10th to 12th place in the list to the third. This is very controversial, but not to take the population density into account, while speaking about networked access, is unacceptable. It is possible that the impact of population density on the overall index should be smaller.

The innovation component was also a proxy calculation since it takes into account only the potential of a country and does not consider how this potential is being utilised at the present time. In our opinion, the results of table 22, as they are now, reflect mostly the potential for achieving e-readiness rather than a real e-readiness.

Conclusion

All the transition and emerging market economies of Eastern Europe and CIS are now permanently connected to the Internet, have some ICT infrastructure and, consequently, some elements of a knowledge-based economy. The Internet is growing at a rate unprecedented in the history of any technology. Yet, significant disparities in access to the Internet exist along the lines of national, regional and population income groups.

The impact of ICTs on these countries is best understood in terms of the three subgroups they form. Those countries that have economic and technical resources to adopt and utilise the ICTs; to exploit their full potential to transform their economies; and to join the group of the most developed European countries, constitute the top, or the highest subgroup.

Those countries that are capable of adopting and adapting over the next few years constitute the middle subgroup, and the third subgroup is composed of those countries that are in danger of being left behind altogether in the ICT process and in the knowledge revolution. The borders between subgroups are not very clear. Only some countries fit exactly into the first subgroup ie. the Czech Republic, Slovenia, and Hungary. Similarly only some countries fit exactly into the second subgroup ie. Bulgaria, Romania, and the Ukraine, while some countries, such as Poland, the Slovak Republic, and the Baltic States have a lot of parameters from the first subgroup, but also of the second subgroup.

Many countries of the former USSR formally belong to the third subgroup, but have some optimistic parameters from the second subgroup. Russia is very specific case. Formal parameters and numbers per inhabitants reflect that Russia is a member of the second subgroup. However, the absolute figures for Internet and telecom penetration in Russia; the very large territory on which these figures are achieved; the size of the population compared to the other countries in transition; the concentration of some countries of Eastern Europe and CIS around Russia; all these, from the point of view of the Internet and the knowledge-based economy, make the Russian Federation one of the main Internet and knowledge-based economy players not only in Europe, but also on a global scale.

The division of all the countries of Eastern Europe and CIS into three subgroups has been carried out not for the purpose of "labelling" them, but in order to understand the main tendencies and trends for the future development of the Internet and the knowledge-based economy in Eastern Europe and CIS. The divisions, therefore, are somewhat "virtual".

The cost of access to the Internet continues to decline, and broadband access with a high quality of service is becoming available to a comparatively large segment of users in the highest subgroup and to many countries in the middle subgroup. However, this drop in costs is yet to translate into affordable access for many other countries of the middle and the lowest subgroups. Proactive policy interventions at an international level (peering and transit arrangements); at a national level (telecommunication tariffs); and at a local level (promoting and providing for collective access); may be necessary, in many instances, to make the Internet accessible to a wider segment of the population.

In addition to the inter-country divide, critical intra-national disparities exist across all countries of Eastern Europe and CIS. Income of individuals/households is a major source of these intra-national disparities, but income also correlates with other dimensions of disparity, such as gender, level of education and location of users. Ultimately, the interaction of these different dimensions is important when evaluating the significance of any one dimension.

Many problems exist for all countries in transition and emerging market economies of the UNECE region, as well as for some countries globally, and this leads to a delay in the development of an information society and a knowledge-based economy. These are as follows:

- ICT service is available for the wealthiest, most socially active and digitally-literate people and societies (a digital divide exists inside almost every country);

- only a few countries in Eastern Europe and CIS have solved the problem of ensuring a mass, free and public access to the modern Internet and telecoms services (a digital divide between countries also exists);

- there is comparatively low PC penetration and, significantly, PC penetration in the education sector is low;

- a strong policy and clear understanding among individuals, economic agents and policy makers of the importance of real support for the development of an information society and the building-up of a knowledge-based economy, exists in only a few countries(a digital divide in minds also exists);

- inefficient coordination of existing electronic information resources in the countries of Eastern Europe and CIS is one of the main obstacles that limits the development of a knowledge-based economy;

- local content of almost every country in transition and emerging market economy is inadequately represented in global networks;

- in some countries with strong scientific and technical potential, such as the Russian Federation, the Republic of Belarus etc., there are no reliable mechanisms for the commercialisation of scientific and technological achievements;

- in many cases, the support of Government and a "law and order regime" are not enough to integrate a certain country into a global knowledge-based economy.

In spite of all the above problems, in many countries of Eastern Europe and CIS, a platform for the development of a knowledge-based economy has been already formed. Over the last five years, ICTs have penetrated all spheres of daily life and work. In some countries in transition and emerging market economies, the ICT market is already liberalised (in the Russian Federation, Czech Republic, Slovak Republic). In some other countries liberalisation is ongoing (Hungary, Latvia), while in others, it is planned for the next year or for the near future (Yugoslavia, Bulgaria). Many countries of Eastern Europe and CIS have enormous and unique information resources. The market of electronic information and services is very dynamic in Eastern Europe and CIS. The Internet audience is estimated to constitute tens of millions of users with 17.5 million registered in 2001. Many countries of Eastern Europe and CIS have a unique potential of highly qualified personnel in the ICT sector, which continues to grow.

On the Eastern European and CIS markets there exists practically all models of electronic business. Governments in many countries have started to implement a range of strategic programmes and initiatives aimed at achieving an effective integration of their countries into the world economy. A very good example of these programmes is Program "Electronic Russia". Large investment in the ICT sector, as well as support from the international community, could significantly accelerate this process.

Bibliography

1. Towards a Knowledge-Based Economy. Country Readiness Assessment Report. Georgia.
2. Towards a Knowledge-Based Economy. Country Readiness Assessment Report. Russian Federation.
3. Towards a Knowledge-Based Economy. Country Readiness Assessment Report. Armenia.
4. Towards a Knowledge-Based Economy. Country Readiness Assessment Report. Bulgaria.
5. Towards a Knowledge-Based Economy. Country Readiness Assessment Report. Kyrgyz Republic.
6. Towards a Knowledge-Based Economy. Country Readiness Assessment Report. Latvia.
7. Towards a Knowledge-Based Economy. Country Readiness Assessment Report. Yugoslavia/Serbia and Montenegro.
8. Towards a Knowledge-Based Economy. Country Readiness Assessment Report. Republic of Belarus.
9. Towards a Knowledge-Based Economy. Country Readiness Assessment Report. Slovak Republic.
10. Materials of ITU, www.itu.org.
11. Materials of OECD, www.oecd.org.
12. Materials of UNECE, www.unece.org.
13. . Materials of World Bank, www.worldbank.org.
14. Thomas A. Stewart Intellectual Capital. The new Wealth of Organizations. Doubleday/Currency, 1997.
15. Materials of the Centre For International Development at Harvard University. The Guide "Capture the Benefits of the Networked World, www.readinessguide.org.
16. Ghilain Robyn, Director Statistics and Information Network Branch of UNIDO. Measuring the Economic Importance of ICT. Discussion Paper No 2, 2001.
17. Materials of TeleGeography Company, www.telegeography.com.
18. Hungary's SuliNet web portal materials, www.sulinet.hu.
19. Hungary's MEDINFO web portal materials, www.medinfo.hu.
20. Tim Kelly, Michael Minges, Lara Srivastava (ITU), and Jozsefne Pergel (Hungary) Internet in a Transition Economy: Hungarian Case Study, 2001, www.itu.org.
21. Nikola Krastev East: Report Says Information Technology Development Lagging. Radio Liberty Materials. www.rferl.org.
22. Materials of Global Internet Policy Initiative. www.gipiproject.org.
23. Materials of World IT Report. Russia and CIS. Eastern Europe. www.worlditreport.com.
24. S. Nanthikesan. Trends in Digital Divide. Harvard Center for Population and Development Studies. Cambridge. MA 02128. 2000.
25. McConnell International and World Information Technology Service Alliance (2000), Risk E-Business: Seizing the Opportunity of Global E-Readiness, www.mcconnellinternational.com.
26. The World Economic Forum Materials. Highlights of the Global Competitiveness Report 2001 – 2002.